Archimedes

NEW STUDIES IN THE HISTORY AND PHILOSOPHY OF SCIENCE AND TECHNOLOGY

VOLUME 57

Archimedes has three fundamental goals: to further the integration of the histories of science and technology with one another; to investigate the technical, social and practical histories of specific developments in science and technology; and finally, where possible and desirable, to bring the histories of science and technology into closer contact with the philosophy of science.

The series is interested in receiving book proposals that treat the history of any of the sciences, ranging from biology through physics, all aspects of the history of technology, broadly construed, as well as historically-engaged philosophy of science or technology. Taken as a whole, Archimedes will be of interest to historians, philosophers, and scientists, as well as to those in business and industry who seek to understand how science and industry have come to be so strongly linked.

Submission / Instructions for Authors and Editors: The series editors aim to make a first decision within one month of submission. In case of a positive first decision the work will be provisionally contracted: the final decision about publication will depend upon the result of the anonymous peer-review of the complete manuscript. The series editors aim to have the work peer-reviewed within 3 months after submission of the complete manuscript.

The series editors discourage the submission of manuscripts that contain reprints of previously published material and of manuscripts that are below 150 printed pages (75,000 words). For inquiries and submission of proposals prospective authors can contact one of the editors:

Please find on the top right side of this webpage a link to our *Book Proposal Form*.

More information about this series at http://www.springer.com/series/5644

Gideon Manning
Editor

Testimonies: States of Mind and States of the Body in the Early Modern Period

With contribution by Lisa Jardine

Springer

Editor
Gideon Manning
Claremont Graduate University
Claremont, CA, USA

Cedars-Sinai Medical Center
Los Angeles, CA, USA

ISSN 1385-0180 ISSN 2215-0064 (electronic)
Archimedes
ISBN 978-3-030-39374-8 ISBN 978-3-030-39375-5 (eBook)
https://doi.org/10.1007/978-3-030-39375-5

This Springer imprint is published by the registered company Springer Nature Switzerland AG.
The registered company address is: Gewerbestrasse 11, 6330 Cham, Switzerland

In Memoriam
Lisa Jardine
(1944–2015)
multum in parvo

Acknowledgments

I wish to acknowledge the patience and resolve of the contributors, the timely advice of Mordechai Feingold, and especially to remember Lisa Jardine, who had faith enough in me to ask that I complete the volume that she had started. An equal expression of thanks is due to Lisa's husband, John Hare, who assured that Lisa's essay would be included here. I extend my final expression of gratitude to my own family—Melissa Pastrana, Colum Manning, and Quinn Manning—whose support makes my intellectual life possible.

December 2018

Gideon Manning

Contents

About the Contributors

Peter R. Anstey is Professor of Philosophy in the School of Philosophical and Historical Inquiry at the University of Sydney. He specializes in early modern philosophy with a focus on John Locke, Robert Boyle, and the French *philosophes*. He is the Author of *John Locke and Natural Philosophy* (Oxford, 2011) and Editor of *The Oxford Handbook of British Philosophy in the Seventeenth Century* (Oxford, 2013).

Sorana Corneanu is Associate Professor of English and Researcher in early modern studies at the University of Bucharest. She is Author of *Regimens of the Mind: Boyle, Locke, and the Early Modern Cultura Animi Tradition* (Chicago: University of Chicago Press, 2011); Editor of "The Care of the Self in Early Modern Philosophy and Science" (a special issue of the *Journal of Early Modern Studies* 4.2, Bucharest: Zeta Books, 2015); and Coeditor, with Guido Giglioni, James Lancaster and Dana Jalobeanu, of *Francis Bacon on Motion and Power* (Dordrecht: Springer, 2016).

James Everest completed his PhD in intellectual history at the University College London (UCL) in 2017 and currently holds teaching fellowships at UCL and the University of Birmingham. A section of his PhD was published in *Intellectual History Review* as "Francis Bacon's Method and the Investigation of Light." He has recently been working on the postwar life of Otto Wächter, the Nazi Governor of Galicia (now western Ukraine), in a project that led to the BBC Radio 4 documentary and podcast *Intrigue: The Ratline*.

Anthony Grafton teaches European history at Princeton University. He studied history, classics, and history of science at the University of Chicago and University College London. His recent publications include *The Culture of Correction in Renaissance Europe* (2011) and (with Richard Calis, Frederic Clark, Christian Flow, Madeline McMahon, and Jennifer Rampling) "Passing the Book: Cultures of Reading in the Winthrop Family."

Anita Guerrini is Horning Professor in the Humanities and Professor of History at Oregon State University (emerita as of 1 January 2019). Her most recent book, *The Courtiers' Anatomists: Animals and Humans in Louis XIV's Paris* (University of Chicago Press, 2015), won the 2018 Pfizer Prize of the History of Science Society for Best Scholarly Book. Her new book project looks at giants, fossil bones, and nationalism in early modern Europe. She continues to work on early modern anatomy and natural history and on the role of history in ecological restoration. She has won grants for her research from (among other agencies) the National Science Foundation, CNRS, NEH, the Max Planck Institute for the History of Science, the Camargo Foundation, and the Descartes Center in Utrecht.

Felicity Henderson is Senior Lecturer in Archives and Material Culture in the Department of English and Film, University of Exeter, having previously been based at the Royal Society as History of Science Events and Exhibitions Manager. She is currently editing Robert Hooke's diary for Oxford University Press and has published a number of articles and book chapters on Hooke and the early Royal Society. Her current research interests include images, translations, manuscript cultures, and narratives of science. She is a Co-investigator on the AHRC-funded project "Making Visible: The Visual and Graphic Practices of the Early Royal Society."

Sarah Hutton is Honorary Visiting Professor at the University of York. She has published extensively on early modern intellectual history, especially on the Cambridge Platonists, and women in early modern philosophy and science. Her publications include *Anne Conway: A Woman Philosopher* (CUP, 2004) and *British Philosophy in the Seventeenth Century* (OUP 2015). She is Director of International Archives of the History of Ideas and Member of the editorial board of the *British Journal for the History of Philosophy* and of the board of management of the *Journal of the History of Philosophy*.

Lisa Jardine was Professor of Renaissance Studies at the University College London where she was also Director of the Centre for Editing Lives and Letters and the Centre for Humanities Interdisciplinary Research Projects. One of Britain's leading historians and public intellectuals, she was a Fellow of the Royal Society and of the Royal Historical Society and an Honorary Fellow of King's College and Jesus College, both in Cambridge. To nonacademics, she may best be remembered for her Radio 4 series "A Point of View" and "The Age of Science" and for her public service as Chair of the Human Fertilization and Embryology Authority. She was the Author of more than 50 articles and 17 books, including *Worldly Goods: A New History of the Renaissance* and *Ingenious Pursuits: Building the Scientific Revolution*. Her book *Going Dutch: How England Plundered Holland's Glory* won the prestigious Cundill International Prize in History. Her most recent book, a collection of essays entitled *Temptation in the Archives*, was published by UCL Press.

Gideon Manning is Research Associate Professor at Claremont Graduate University and Visiting Associate Professor at the Cedars-Sinai Medical Center, having previously taught at the University of Pittsburgh and served on the faculties of the College of William and Mary and the California Institute of Technology. His recent publications include "Health in the Early Modern Philosophical Tradition" (2018) and "Descartes and Medicine" (2019). He is one of the editors of *Professors, Physicians and Practices in the History of Medicine: Essays in Honor of Nancy Siraisi* (2017). At present, he is engaged in a number of projects trying to productively bring together the history of medicine, science, and philosophy.

Anna Marie Roos is Professor of the History of Science and Medicine at the University of Lincoln (UK) and a Fellow of the Linnean Society and the Society of Antiquaries of London. Her recent publications include *Martin Lister and His Remarkable Daughters: The Art of Science in the Seventeenth Century* (2018); *Web of Nature: Martin Lister (1639–1712), the First Arachnologist* (2011); and *The Salt of the Earth: Natural Philosophy, Medicine, and Chymistry in England, 1650–1750* (2007). Her next book, a cultural history of the goldfish, will be published with Reaktion Press in 2019.

Chapter 1
Editor's Introduction

Gideon Manning

Keywords Lisa Jardine · Francis Bacon Foundation

The most important thing to know about this volume is that it was originally conceived by Lisa Jardine, initially organized by her, yet left unfinished by her death. Although the contributors and I have worked hard to realize her vision, everyone would agree Lisa would have improved the resulting publication in nearly every way. My involvement first began as a co-editor in the Summer of 2015, little over a year after Lisa had been Visiting Distinguished Professor at the California Institute of Technology, where I was at the time an assistant professor of philosophy. Lisa's visit, though short, lasting only 10 weeks, was a memorable one for my colleagues and me, overlapping, as it did, with moments of high academic drama and intrigue. Lisa was one of the bright spots for us during that period.

She came to Caltech as a result of winning the 2012 Francis Bacon Award, an honor she received for her record of outstanding scholarship and the substantial impact she and her work had had on the history of science, the history of technology, and historically-engaged philosophy of science. In addition to receiving a medal with Bacon's likeness and a cash prize, the Bacon Award obligated Lisa to be in residence at Caltech, teach a course to undergraduates, deliver a public lecture (she spoke about the career of her father, the British mathematician and historian Jacob Bronowski), and convene a conference on a subject of her choosing. This volume originates from Lisa's Bacon conference. In her own words, taken from the opening remarks of her lecture at the conference and included below as Chap. 3, she planned to use the conference "to reconnect health and thought in the early modern period." As Lisa well understood, health and thought were treated together in the

G. Manning (✉)
Claremont Graduate University, Claremont, CA, USA

Cedars-Sinai Medical Center, Los Angeles, CA, USA
e-mail: gideon.manning@cgu.edu

G. Manning (ed.), *Testimonies: States of Mind and States of the Body in the Early Modern Period*, Archimedes 57,
https://doi.org/10.1007/978-3-030-39375-5_1

seventeenth century and she believed that reuniting them in our scholarship would "add a significant dimension to our historical understanding."

Lisa was the one who drew up the list of speakers—as usual with her, the list was diverse and included younger scholars whose work she recognized and virtuously championed as well as practiced hands in the history of science and philosophy. Ultimately she presided over a conference titled "Testimonies: States of Mind and States of the Body in the Early Modern Period" that included lectures given sequentially by James Everest, Lisa Jardine, Felicity Henderson, Peter Anstey, Rob Iliffe, Karin Ekholm, Evan Ragland, Gideon Manning, Sarah Hutton, and Anita Guerrini.

When Lisa died in October 2015, her Bacon volume necessarily changed. After a conversation between Jed Buchwald, Mordechai Feingold and myself, it was decided that the volume would now be a memorial volume in Lisa's honor. With this change two decisions followed. First, we decided to extend the length of time we would allow for people to contribute while also extending invitations to additional authors. Second, we decided to contact Anthony Grafton, Lisa's longtime friend and collaborator, to ask if he would contribute a memorial piece in Lisa's honor. Of the original attendees at the conference, Rob Iliffe, Karin Ekholm, and Evan Ragland were unable to contribute to the present volume, and my own contribution has been limited to this introduction and my editorship, but Anna Marie Roos and Sorana Corneau both gratuitously agreed to join us, as did Anthony Grafton.

In working to shepherd Lisa's Bacon volume to press, I have tried to do as little as possible to stand in the way of her hopes for the conference and the volume she began to edit. I have tried, for example, to provide the authors the same freedom Lisa intended them to have, even when this has meant some inconsistency among their bibliographies; some contributors chose to list collected works by the original author, while others list the same work by the subsequent editor because they discuss those editors' choices in their essay. Even so, I am not certain how Lisa would have handled this inconsistency or whether she would have written an expansive introduction. Had she done so, she likely would have elaborated on her goal of reconnecting health and thought, for she had an even more specific target that comes into clear focus in her essay, namely, attitudes toward mind-body interaction as revealed in acts of self-diagnosis and experimentation. Again, citing her own words, Lisa believed by studying "the way self-experiment, combined with careful attention to the states of mind which accompany states of body specifically altered by measured doses of chemically prepared substances… [we can discover] a new and influential attitude to body/mind interactions" in the seventeenth century. She had a "further suggestion" as well, "that in the surviving records of such experimenting on one's own body, we can observe individuals like Bacon, Boyle and Hooke, deliberately setting out to repeat pleasurable, or intellectually productive moods and states of mind, by applying the same 'physic' [medicine] on successive occasions (often, daily)." In this we can witness "therefore, theories of the working of the human mind being developed by key members of an urban culture (London; interregnum Oxford), who based those theories on their own regular, long-term use of self-administered, mind-altering substances." In my words now, there was nothing accidental about Bacon, Boyle or Hooke's efforts to boost their intellectual activity and control their bodily functions. It is hardly an overstatement to claim that there was a significant drug culture in the early modern period linked to self-experimentation, new medicines, and the new science. This is one of the many things this volume, and Lisa's final published work, has to teach us.

* * *

The chapter after this introduction is Anthony Grafton's. Although his contribution shares some of his personal knowledge of Lisa, he also provides those of us who never collaborated with her the opportunity to peer over the shoulder of two virtuosi working together, and to witness Lisa thinking as we appreciate the impact she had even on those most gifted of specialists who found themselves in her company. Grafton recounts their shared sense that "the explication of texts formed the core of classroom activity" in the sixteenth century and the way this shared sense led to their joint work on *From Humanism to the Humanities*. He recalls their process of writing together and the state of intellectual history in the 1970s and 1980s that prompted them to discover the intellectual and tactile experience of reading in a prior age through an examination of Gideon Harvey's personal library held in Princeton. Together they came to see what Harvey "made of individual books, but also to situate him both as a reader," for his "marginalia illuminated the material conditions within which he worked," and a "facilitator," for he guided his contemporaries' interaction between ancient texts and their political present, avoiding the "erudite tradition of philological analysis" in favor of a thoughtful hermeneutics meant to teach the gentleman class. Grafton calls forth two lessons Lisa would surely agree with from their time together, but the second is the more informative for where it led Lisa later on. Grafton believes there was something deeply contingent about their success: Harvey's books were not open to all and they were fortunate to find them not just because that they had been preserved and donated to Princeton but because they (Lisa and Tony) had the rarefied credentials to view them. So keenly aware of this contingency was Lisa that, as Grafton recounts, she worked to end it with efforts to digitalize Harvey's work in her Archeology of Reading project. This is just one instance, among many, where an idea that took root in Lisa's scholarship grew to bear fruit for the rest of us.

Lisa's contribution, still very much as she delivered it at the original Bacon conference, comes next as Chap. 3. Opening with some remarks about the Bacon conference, Lisa moves quickly to the themes mentioned above. She does this by anchoring her discussion with two starting points. The first is Robert Hooke's effort to conduct "trials" with chemical compounds, using them as instruments for enhancing his mental capacities just as he used the microscope and telescope to enhance his vision. The second starting point is Francis Bacon's personal purging routine, the details of which appear to have been suppressed by Bacon's later editors. As Lisa explains, his regimen of 'physic' aimed at prolonging his life and entailed awareness of and a manifest interest in the effects of chemical medicine on the mind and body. Although not all medical men of the period were so enthusiastic for the educated gentleman's self-prescription of purgatives and toxic substances, Bacon and many others continued self-experimentation up to the time of their death. One of the many larger lessons to draw from these case studies linking health and thought is that interest in experimentation in the early modern period was, often, fundamentally concerned with the relation between mind and body. Indeed, there is a strong link joining early modern science and philosophy to experimental medical activity where theories of the human mind were being developed.

James Everest provides the next contribution, pressing further into the themes broached in Chap. 3 to tell a fuller story of Bacon's interest in longevity and his own health. Picking up on the connection between thought and feeling, Everest's Bacon comes into focus through episodes of ill health and an acute awareness of his own suffering. As Everest presents Bacon, he was well aware of what it was like to be let down by his body. Further, Bacon's experience of being embodied is, claims Everest, an essential backdrop for an important strand of his natural philosophy: his writing on life and death. But beyond Bacon's choice of natural philosophical subject-matter, Everest suggests Bacon's experience of his own embodiment may also have played a role in his approach to natural philosophical questions more generally. While Bacon advocated the advancement of learning by means of information gained from experiments he also pursued his own sophisticated course of experimental investigations, the latter of which have been marginalized in the retelling of Bacon's natural philosophy. Everest, however, connects the two in his chapter, suggesting that the value Bacon placed on experimental knowledge may have owed something to his life-long practice of and perceived benefit from medical self-experimentation.

Sarah Hutton's contribution in Chap. 5 shifts our attention from prolonging life to the experience of pain and trying to relieve it. Here she continues her groundbreaking work on Anne Conway, who suffered debilitating pain throughout her life. Sutton discusses several interconnections between explanations of pain and the experience of pain in seventeenth-century England by focusing on an unusually well-documented episode in the history of pain relief: the visit to England in 1665 by the Irish healer, Valentine Greatrakes. This is an episode which captures the complexity of the early modern medico-philosophical setting and the human experience of illness. She further considers the role of Philip Stubbe and discusses the role of pain in Anne Conway's philosophy to argue that her philosophical treatise reflects her acquaintance with medical theories which were in circulation at the time of the Greatrakes affair.

Felicity Henderson's chapter comes next and details Robert Hooke's account of memory. Examining an unusual lecture contained in Hooke's *Posthumous Works*, one that begins with the subject of light but quickly turns to outline a theory of mind that ranges over topics such as memory and, to a lesser extent, the perception of time, and the action of the soul, Henderson makes a case for viewing Hooke's account of memory as productively combining elements from his work on microscopy, astronomy and physics, even offering a new perspective on Hooke's geophysics. Henderson goes on to make an especially provocative and personal claim about Hooke. Noting that his lecture on memory should be seen in connection with his efforts to discuss an explicit scientific method, what he called his "philosophical algebra," and that his theory of mind can be seen to emerge directly from his own experimental practices, Henderson believes we can gain privilaged insight into

Hooke's own working mind or, at minimum, Hooke's vision of the ideal mind of an early-modern polymath by examining Hooke's account of memory.

Anita Guerinni provides the next contribution, turning our attention to the shared mental capacities and physical analogies between humans and animals. She asks, in particular, how the widespread and lasting assumption of an analogy between animal and human bodies affects ideas about the mind-body relationship? And how did seventeenth-century men of science justify their employment of animal brains to talk about human brains and minds? Guerinni shows how Galen's ideas of the structure of the brain, based on animals, gave way to a multiplicity of ideas about the relationship between the brain and the mind (or soul), though she also recognizes that the meanings of both these terms—mind and soul—shifted continuously during this era. In addition, she documents the continued reliance on animal models, noting that is seems to conflict with a new emphasis on specifically human structures and behaviors. But at the same time anatomists and philosophers also looked anew at animal cognition. The work of Marin Cureau de la Chambre, Niels Stensen, and Thomas Willis reveals some of these new ideas and the tensions that result.

Peter Anstey's contribution is next in Chap. 8. Adding to his already significant work on John Locke, Anstey's here takes up the theme of self-diagnosis and prescription in Locke's correspondence and published work. He documents how Locke diagnosed not just the nature of the self as a philosophical category with its implications for mental health, Locke also diagnosed the medical states of his own body, showing himself well aware that the mind and body are intimately related. Outlining a number of possible ways to pursue his topic, Anstey choses to develop Locke's philosophy of the self. He specifically considers its implications for the cultivation of health and, beyond that, to determining the body's appropriate place in nature. Traditional philosophical topics emerge in Anstey's chapter—e.g., issues of identity and individuation and the spatial extension of thought—but throughout Anstey's goal is to locate Locke's developing views of the self, the "person" as Locke came to call it, in the context of Locke's broader interests in the correct (healthy) conduct of the understanding. There were, Anstey concludes, myriad ways in which Locke "accentuated the opportunities" for informed self-diagnosis available to him. Indeed, Locke was one of the periods most accomplished diagnosticians of the philosophical and medical self.

The role of the imagination and how it relates to the art of thinking is the subject of the next chapter by Sorana Coreau. Dedicated to Josep Glanvill, who in *The Vanity of Dogmatizing* (1661) and its subsequent versions (1665, 1676) sketched elements of an art of thinking in which the functions of the imagination played a central role, Coreau's contribution finds at the root of Glanvill's reflections a defense of the immortality of the soul, carried out via an attack on faulty modes of thinking. Although the end-result was an extended analysis of erroneous thinking, Coreau characterizes it more fully as an Academic skeptical praise of free inquiry. Not only does Coreau reconstruct this line of Glanvill's thought, she locates his sources, the disciplinary rearrangements it involves as well as the tensions it creates. Ultimately, Coreau defends several claims, including: the art of thinking in question here takes the form of a regulation of defective mental operations and is thus, contrary to a traditional logic, subsequent to, and dependent on, an analysis of error; that this analysis of error is a hybrid development, in which logical categories become mixed with psychological and medical notions; and that it is within the framework of his art of thinking that Glanvill develops his position relative to the correct method of inquiry in natural philosophy.

The volume's final contribution comes from Anna Marie Roos, who returns us explicitly to the issue of self-diagnosis as it appears in the lives and writing of case histories among leading figures of the Royal Society. Specifically discussing the cases of John Wilkins, Martin Lister, and John Ray, Roos documents how personal experiences of disease and pain were shared and objectively documented between and among physicians and natural philosophers. Her examples reveal an expanding vocabulary of objectification of the body, self-experimentation, and an increasing precision of empirical description inspired, the evidence suggests, by Baconian models developed by the early members of the Royal Society. If Roos is correct, and she surely is, we ought to conclude that England's early modern physicians and natural philosophers sought out a language to describe the body and its pains beginning with themselves and their own experiences with disease and dysfunction.

* * *

To those of us lucky enough to have known her, Lisa Jardine was a much beloved scholar and human being, especially admired for the many responsibilities she assumed outside of the academy, both in politics and in Britain's intellectual life. In spite these responsibilities, she continued her dauntingly original research and never ceased to give back to our profession by supporting the careers of younger scholars. This part of her legacy lives on in The Royal Society's "Lisa Jardine Grant Scheme" (https://royalsociety.org/grants-schemes-awards/grants/lisa-jardine/), which is a testament to the immensely positive influence she had, the good will she created, and the highest of standards that she maintained. I sincerely hope she would have approved of the volume we finished for her.

Chapter 2
Lisa Jardine: A Life in the Margins

Anthony Grafton

Abstract This chapter records part of the history of Lisa Jardine's pathbreaking work on Renaissance marginalia and the readers who made them. It is also meant to enable readers who did not have the chance to collaborate with Lisa to witness Lisa at work, and to grasp why joining her in a project had a transformative effect on senior as well as junior scholars. In the 1970s, Jardine and Grafton agreed that the explication of texts formed the core of classroom activity in the sixteenth century: this shared sense led to their joint work on *From Humanism to the Humanities*. They became experienced at writing together, and over time in the 1970s and 1980s, as they continued to discuss the state of intellectual history, they came into contact with the new history of books being created by Robert Darnton, Roger Chartier and others. Conversations with Darnton prompted them to examine books from Gideon Harvey's personal library held in Princeton, and by doing so they discovered the intellectual and tactile experience of reading in a prior age. Together they learned not only to see how Harvey interpreted individual books, but also to situate him as an example of a category that modern scholarship had forgotten, the professional reader. His marginalia revealed much about the material conditions in which he worked. They also showed that he had served as an intellectual facilitator, who guided his contemporaries' interaction between ancient texts and their political present. He stressed that they had deliberately avoided the erudite tradition of philological analysis in favor of a thoughtful hermeneutics meant to draw practical lessons for the present from the past. Jardine's and Grafton's arguments, soon published, inspired both objections and applications (more of the latter than of the former), and their joint article on Harvey became a standard reference for historians of books and reading. Over the years, they modified many of their initial conclusions. They also became aware that Harvey's books had not been open to all: they were able to work on these extraordinary documents not just because that they had been preserved and donated to Princeton but because they (Jardine and Grafton) had the rarefied credentials to view them. So keenly aware of this contingency was Jardine

A. Grafton (✉)
Princeton University, Princeton, NJ, USA
e-mail: grafton@Princeton.EDU

G. Manning (ed.), *Testimonies: States of Mind and States of the Body in the Early Modern Period*, Archimedes 57,
https://doi.org/10.1007/978-3-030-39375-5_2

that she worked to end it with efforts to digitalize Harvey's work, among others. These eventuated in the Archeology of Reading project that she undertook in collaboration with Earle Havens at Johns Hopkins and her colleagues at CELL, and which has now made a substantial number of books annotated by Harvey and John Dee available on line, with searchable transcriptions and translations. This is just one instance, among many, where an idea that took root in Lisa's scholarship grew to bear fruit for what her subjects would have called the Republic of Letters.

Keywords Lisa Jardine · Gabriel Harvey · Rober Darnton · Lucius Wilmerding · History of books · History of reading · Marginalia · *Archaeology of Reading*

The history of books and readers was created at Princeton at the end of the 1980s. Lisa Jardine did the most to bring it into being, and I had the good fortune to be present at the creation, and to collaborate with her. We had first met long before, at the Warburg Institute, where Lisa held a Senior Research Fellowship from 1971 to 1974. She was immensely impressive. Erudite, articulate and confident, she showed no tolerance for weak arguments or imprecise readings of sources. Her seminar talks were events. Chance sent both of us to spend the academic year 1974–1975 at Cornell, where we attended one another's seminars and began to collaborate.

Both of us were already working with printed books with marginalia when we met in London, and we discussed them while at Cornell. Lisa—who had studied mathematics and English at Cambridge and translation at Essex—devoted her Cambridge doctoral dissertation to tracing the connections between Francis Bacon's program for the reform of learning and the forms of dialectic that flourished in his youth. She had examined the annotations that the Cambridge scholar Gabriel Harvey left in works by Cicero and Quintilian as part of her wider inquiry into the ways in which sixteenth-century humanists taught their students the arts of reasoning and literary analysis. Meticulously trained by Robert Bolgar, she brought to this study a close attention to revealing details of every kind, from the ways in which marginalia were laid out on the page to the words and symbols used in them.

Harvey had long served literary historians and lecturers as a figure of fun. They took their cue from Harvey's own contemporaries. Edward Forsett satirized his penchant for speaking in polysyllables and bragging about his marginalia in a Cambridge student play, *Pedantius*. Worse still, Thomas Nashe, who possessed the sharpest nib in Elizabethan London, had stabbed Harvey with it repeatedly. His polemical tracts against Harvey etched an unforgettable portrait of a scholarly Malvolio, self-obsessed and ridiculous. For Nashe, everything from Harvey's dress to his "flourishing flantitanting gouty Omega fist," revealed his empty vanity.[1] Reading out bits of Nashe's *Have With you to Saffron Walden* could enliven a dull lecture on Elizabethan prose or literary politics.

[1] See Hutson 1989.

Lisa had little interest in, or patience for, this tradition of scholarship. But Harvey and his marginalia had long benefited from a different kind of attention as well. A long note in his copy of Thomas Speght's 1598 edition of Chaucer, preserved in the British Library, included what remains the earliest known reference to Shakespeare's Hamlet: "The younger sort takes much delight in Shakespeare's Venus, and Adonis: but his Lucrece, & his tragedy of Hamlet, Prince of Denmark, have it in them, to please the wiser sort."[2] Other marginalia revealed that Harvey was very close to Edmund Spenser. By the 1970s, specialists in Elizabethan and Jacobean literature had been quarrying nuggets like these from the margins of Harvey's books for decades. As early as 1913, G.C. Moore Smith, a Berlin-trained Sheffield professor of English, published a rich selection from Harvey's annotations with A.H. Bullen's Shakespeare Head Press.[3] Though many of Harvey's other books came to rest in the British Library or in Oxford or Cambridge collections, a fair number remained in the market in the decades around 1900, when American collectors took a special interest in the history and literature of the Elizabethan and Jacobean periods. These were soon scattered across private and public collections, from Princeton, New Jersey, to San Marino, California. Some of them attracted serious study.[4] By the 1970s, Virginia Stern, an erudite and energetic private scholar, was known to be at work on an inventory and study of Harvey's books and marginalia. She produced the first modern survey of his library in 1972.[5]

Lisa agreed with Stern—and Moore Smith before her—that Harvey's annotations recorded the interests and ideas of a scholar who deserved to be taken seriously. But she insisted that still more—much more—needed to be done. To study Harvey's marginalia properly, one had to read the texts in which they appeared as well as the notes themselves. And to do that—in the case of Harvey's notes on ancient and modern works on rhetoric and dialectic—the modern scholar needed a full training in the traditions of rhetoric and dialectic, from their origins in Greece to the forms that flourished in the Cambridge curriculum in the sixteenth century. As the working notes of an active teacher of rhetoric, Lisa saw, Harvey's annotations could shed a brilliant light on the ways in which Francis Bacon and many others had been taught to formulate arguments. No one had analyzed them in anything like this depth before.

I had also started exploring the words inscribed in margins long before. In 1974, hard at work on my doctoral dissertation on Joseph Scaliger, I found myself confronted by a classical question in the history of science: how to know if contemporaries would have found a bold assertion plausible. Scaliger devoted himself to a study that now invites the adjective "mere," but which won him the only full-time

[2] British Library Add MS 42518, fol. 422^{v}. For a reproduction of the original page and a fuller transcription, see https://shakespearedocumented.folger.edu/exhibition/document/manuscript-marginalia-gabriel-harvey-refers-hamlet-lucrece-and-venus-and-adonis (consulted November 23, 2018). On the dating of Harvey's notes, see Hirrel 2012.

[3] Harvey 1913.

[4] Bourland 1940; Wilson 1948.

[5] Stern 1972. Stern would develop this much further in Stern 1979.

research professorship between the fall of the Alexandrian Museum and the rise of the Institute for Advanced Study: chronology, the analysis of calendars and dates. His last massive work on the chronology of the ancient world, the *Thesaurus temporum*, appeared in 1606. In it he published, among a great deal else, the list of Egyptian dynasties compiled in Greek, in the third century BCE, by a priest named Manetho, from Sebennytos in Lower Egypt. This list had one starkly visible disadvantage. Adding up the lengths of all the reigns listed in Manetho's 31 dynasties showed that Egyptian history had begun not only before the universal flood, but also before the Creation. Scaliger, a committed Calvinist, knew that the Bible preserved the true history of the world. Pagan claims that it was far older than the Mosaic account suggested must be false. Still, he found himself unable to reject the list as a forgery. It seemed too strange to be the product of a Hellenistic historian inventing a deep past. So he printed it. To make it fit into a historical framework, he assumed a period of what he called "proleptic time"—time that ran before there were a sun and a moon to generate years and days.

When Egyptian hieroglyphs were deciphered, long after Scaliger's death, his conjecture was verified. But was it a plausible argument, in period terms? I wondered how Casaubon—one of the two or three contemporaries whose scholarship Scaliger took seriously—would have responded to it. I suspected he would have disapproved, but I could not prove it. At this point, a solution materialized. I held a Fulbright scholarship at University College London. A fellow Fulbright scholar at Cambridge told me that the University Library had systematically collected books with marginal annotations by great scholars, including Casaubon. In the mid-nineteenth century, Luard and Babington had even drawn up a catalogue of them—a slim volume that one could buy, in those days, for something like 45 pence.[6]

Luard's list informed me that the Cambridge Library owned Casaubon's copy of Scaliger's *Thesaurus*. More remarkable still, the book answered my question. Casaubon dismissed Scaliger's discussion of Manetho and proleptic time: "I don't see what use these inventions of foolish peoples are for true history." The annotation transmuted my conjecture into fact. From then on, my own practice as a scholar involved a great deal of searching for marginal notes that could reveal how readers had responded to technical books.[7]

During the year that Lisa and I both spent at Cornell, we began to discuss humanist education—a subject both of us addressed in our seminars. We did not believe in the conventional account, which held that the humanist classroom had been a place of open intellectual inquiry, where students were not only initiated into the study of Latin and Greek, but also given complex lessons in ethics and history. These descriptions were drawn mainly from the manifestoes of humanist teachers and the lectures they gave on the opening days of their courses—neither of which could be seen as transcripts of day-to-day teaching. We began to approach the subject from a new direction. In early modern schools and universities, the explication of texts formed

[6] Luard and Babington 1864; see Sherman 1995, chap. 1.

[7] For a fuller account see Grafton 1983–1993, II.

the core of classroom activity. Close reading of surviving line-by-line notes, we thought, could bring the historian right into the classroom, to stand by the teacher and appreciate the full richness and the full pathos of past teaching. The result of these discussions was the research that went into *From Humanism to the Humanities* (1986)—a book that shocked traditionalists by its willingness to acknowledge that classroom learning in the Renaissance was not always the garden of earthly delights that humanist pedagogues advertised in their opening lectures. Lisa's readings of Harvey's marginalia formed one of the most provocative sections of a provocative book.

By 1986 we had established an effective routine for collaboration. Vivid memories remain: Lisa, very pregnant, lying on the camp bed in our town house in Headington, arguing about one of the chapters in what was slowly becoming *From Humanism to the Humanities*; Lisa, designating me as the drinker when we finally handed in our typescript to Colin Haycraft, so that I had to be scraped up from a chair in his garden and sent home in disgrace in a taxi; above all, Lisa and me, writing together, one of us dictating and the other typing, and the one who typed paying no attention to the one who dictated. This worked very well so long as we changed places from time to time. We learned that collaboration offered distinctive rewards, as each of us pushed the other in directions previously undiscovered. Perhaps we were already primed to work on Harvey.

But our move to a history of reading also had a wider context. In these years, intellectual history entered a long crisis. One of those who did much, first to cause the crisis and then to resolve it, was Robert Darnton, a brilliant historian of the eighteenth century based at Princeton. Through the late 1960s and early 1970s, he engaged in a campaign against what he described as the "armchair" methods of traditional intellectual historians like Peter Gay. Specialists on the Enlightenment had devoted themselves to reading the texts of writers acknowledged to be the intellectual leaders of their time: Voltaire, Diderot, Rousseau. How, Darnton asked, could they know if these writers had in fact met with any response from their contemporaries? And what could they know of the ways in which they were read?

Instead of simply reading the works that had been entombed in massive leather-bound editions, he argued, historians must treat the French culture of the Enlightenment as a living world that had existed in three dimensions and living color. They must recreate the complex Rube Goldberg publishing world of the time and identify those texts, and those editions of texts, that actually reached a large public. And to do this they must investigate every conceivably relevant record, from the notes of censors and police spies in Paris to the records of the great Neuchâtel publishing house that became Darnton's single greatest source of material. This new program of research created a new vision of Enlightenment culture as a working system of communication that linked not only authors, publishers and readers, but also spies, informers and pamphleteers.

Yet by Darnton's own admission, his program for reconstructing the system of textual production and consumption fell short at one crucial point. He did not know how to recreate the intellectual and tactile experience of reading: how to find out what his subjects thought and felt as they went through the pornographic novels which—as he taught us—formed as effective a vehicle for Enlightened thought as the raciest polemics and fictions of Voltaire and Diderot (the meaning of which to readers remained equally obscure). Darnton declared the difficulty with absolute frankness. Where his methodological articles on the nature and impact of publishing were sharply formulated, his first discussion of how books were actually consumed bore an uncharacteristically tentative title: "First steps towards a history of reading," and asked an uncharacteristically tentative question: "Reading has a history. How can we recover it?"[8]

With all the bristly helpfulness of youth, I showed Darnton a long and labyrinthine article, in which I traced, in detail, how Scaliger had responded to a learned contemporary book on papyrus. He had filled its margins not only with marks and summaries, but also with crisply formulated Latin remarks and criticisms, which he then worked up into a formal critique.[9] This, I argued, was a reader's direct response, recorded and accessible. No, Darnton replied, this was not reading, it was something else. It was understandable that he did not share my interest in an erudite humanist's scrawled insults in Latin. But I could not accept his dismissal of what seemed potentially rich materials. I did not have a counter-argument, but somehow, I thought, the answer, or part of it, must be here.

Darnton and I left our discussion there, but it continued to worry me. It also worried Lisa, when she arrived in Princeton in 1988 to spend a semester as a fellow of the Shelby Cullom Davis Center for Historical Research in my department. The Center had invited Lisa to pursue her research on a different side of book history: how Erasmus and other humanists had managed their reputations in print. The time she spent in Princeton, then and later, yielded a brilliant article, and enabled her to lay the foundations for an original and influential book.[10] Yet Darnton's question would not let her loose. And she suspected that a way to answer his question might be surprisingly near at hand. Several of Harvey's books, owned by a wealthy bibliophilic family, were on deposit in the Princeton University Library. Lisa had used them during a quick visit to Princeton in 1974. There, she thought, we might find something like a vivid record of a reader at work.

Lisa and I persuaded the owner, a charming gentleman named Lucius Wilmerding, to allow us to examine his books. We turned first to the grandest of them: Harvey's copy of the Roman historian Livy. From the first day when we opened its shabby leather covers, we found ourselves exploring the ancient jungles and ruined labyrinths of a lost continent. The book itself was an extraordinary document, wreathed in annotation like a Gothic folly covered with ivy. Every page held revelations. We spent day after day in Princeton's Rare Book Room, copying Harvey's notes in pencil, in our neatest scripts, and arguing fiercely over whether to preserve his abbreviations and other conventions (Lisa did, I did not; she was right). Loose-leaf pages piled up. So did Xeroxes. Every time we identified a book that Harvey was

[8] Darnton 1986.

[9] Grafton 1979.

[10] Jardine 1990; see further Jardine 1993, esp. chap. 3.

working on, we would head to EEBO's lineal ancestor, the STC microfilm collection, and make a copy flow. Everything had to be copied so that both of us had all the same material in front of us.

Primitive accumulation was fun. But it was only the beginning. We had to figure out what Harvey was actually doing. Each new interpretation of a note began the same way. We sat in the department's lounge, elegant as a bus depot. One of us would make a suggestion. The other would refute it. Argument would break out, voices would rise, hands would fly. Nervous faces would appear at the lounge door: were Lisa and Tony quarreling? It was only Ashkenazic Jewish love of argument, still a little scary in golden Presbyterian Princeton. Half an hour later we would have agreed on a formulation and be sitting happily again, drinking dreadful departmental tea and gossiping.

Some of our finds were technical: for example, the simple confirmation that in this case, at least, annotation explicitly recorded experiences of reading. Again and again we saw that Livy, for Harvey, was not an absolute authority, but one text among many, ancient and modern, which he continually compared and contrasted. At first, we were baffled by his ability to muster so much comparative material. But then we came across what has now become a famous image: Ramelli's book wheel, the Ferris wheel for books. At one and the same time we understood why this magnificent piece of equipment had been invented, and what sort of reading it facilitated.

Renaissance readers faced an unparalleled range of texts that claimed some sort of authority: ancient and modern, Latin and vernacular. The new bibliographies composed in sixteenth century by Conrad Gesner and others made it possible to gain knowledge about them (enough for some). But the book wheel was a period tool that seemed to embody a period style of reading: a dynamic effort to confront and learn from the vast diversity of sources and information for sale in Harvey's time in the shops in St Paul's churchyard. Studying Harvey's notes, in other words, enabled us not only to see what he made of individual books, but also to situate him as a reader. His marginalia illuminated the material conditions within which he worked and traced connections between them and the intellectual work that he did—a form of intellectual history relatively new at the time, though Peter Burke, Ann Blair, Noel Malcolm and others have carried it much further in the intervening decades—and one that speaks to all of us who have lived through the recent transformation in our own working conditions.[11]

Harvey's notes also situated his form of reading in another way: as work done in a particular social milieu. His comments made clear that he read Livy not in solitude, as Machiavelli had, but in company. He went through the first three books, which dealt with the founding of Rome, with Sir Philip Sidney, before Sidney went on his embassy to the new Holy Roman Emperor, Rudolf II. He traversed the ten books on Hannibal and Rome, a model history of brutal conquest, with Thomas Smith jr., before that unhappy young man traveled to the Irish Ards to carry out his father's plans for pacifying the Irish, but was killed by one of his own men. At times,

[11] See especially Blair 2010.

Harvey found himself taking part in debates that involved some of the most influential political figures of his time. In 1570–1571 Harvey visited Hill Hall, Theydon Mount, the splendid country house of the elder Thomas Smith, Elizabeth's ambassador to France. This was the first country house in England to be decorated with frescoes representing classical myths and histories (since Lisa had not yet made the material turn, as she would a few years later, it did not occur to us to connect the content of these debates with their setting).

Here Smith and others staged a public debate on Roman history:

> Thomas Smith junior and Sir Humphrey Gilbert [debated] for Marcellus, Thomas Smith senior and Doctor Walter Haddon for Fabius Maximus, before an audience at Hill Hall consisting at that time of myself, John Wood, and several others of gentle birth. At length the son and Sir Humphrey yielded to the gentle secretary: perhaps Marcellus yielded to Fabius.

Harvey's patrons and friends, we now realized, belonged to the Elizabethan war party headed by Leicester. Evidently they accepted him as a political counselor, and saw the form of reading he practiced as a serious form of political advice. Harvey's ability to read ancient history, with others, in an insightful and informative way, won him what we came to see as a defined position as a reader, one that other scholars also held in noble families. Contemporaries called such persons "discoursers." We called Harvey a "facilitator,' someone who negotiated the complex interaction of ancient texts and a dangerous present.

At this point, Lisa and I believed, we had established something quite surprising: that reading the ancients with a learned companion not only formed the core of elite education, but continued to form part of elite homosociality in the later Renaissance. Here indeed was an experience of reading: oral, collaborative, and goal-oriented, with a professional specialist guiding men of higher standing through the text. Harvey's readings were shaped, from the outset, by informed and sophisticated decisions about what questions to ask and what aids to employ. Faced with thickets of erudite commentary on Roman history and antiquarian treatises, Harvey and his friends deliberately chose—as he stated—not to enter them at all. They ignored the erudite tradition of philological analysis with which scholars had long identified Renaissance humanism and concentrated on the politically active individuals whose stories Livy told. Traditionally, histories of interpretation in the early modern period have had little to do with humanistic scholarship. Hermeneutics—formal theories of interpretation—have been the province, for the most part, of Reformation scholars, who studied debates about how to read the Bible. Harvey showed us that he, as a scholar working on a pagan text, always set up a kind of interpretative key before he began to read. Informed by his example, we found other versions of this approach elsewhere.

The Davis Center had attracted a brilliant group of fellows, which included David Hall, Donald Kelley and Valerie Flint. The weekly seminars, led by Lawrence Stone at his terrifying best ("there's nothing wrong with your paper but the style and the content"), were always lively and sometimes tempestuous. We watched Inga Clendinnen brilliantly teasing out the meanings of the fall of Tenochtitlan, Lucette

Valensi deftly separating the layers of prejudice and assumption that shaped western accounts of the Ottomans, and Lisa offering the first results of her work on humanists and print. Graduate students, astonishingly clever and unfailingly critical, asked hard questions and poked fun at the grown-ups. Parties were even livelier: no one who saw Lisa and Val do the Cheltenham Ladies College Stomp will ever forget it.

Lisa and I circulated a draft analysis of Harvey's practices as a reader to a special session of this daunting group. Wearing identical neckties, we presided over a wide-ranging critical discussion. Not all of our hearers—who included Peter Brown, Natalie Zemon Davis and Rachel Weil, as well as Darnton and Stone—agreed with our conclusions, but most thought that we should pursue them further. We sent the revised text to *Representations*—which rapidly turned it down as insufficiently self-reflexive—and then to *Past & Present*, where it found a home, and where it has since been read and cited dozens of times.[12] The best forms of discovery in history resemble nothing so much as Alice's fall down the rabbit hole. Harvey's story was one of these: it opened up the history of reading, but it also moved that of political thought and practice and that of learned interpretation in new directions—as a flood of subsequent publications and discoveries has confirmed.[13]

Our research succeeded for two reasons. One was substantive: we did not have a hammer and start looking everywhere for nails. Lisa and I came to Harvey to interrogate him about the connections between reading and intellectual history, and we did our best to make that clear in the article in which we laid out our findings. In other words: the method we developed worked not because it was new, or because it was a method, though new things and methods may be quite good. It worked because we had formulated a historical question, thanks to Robert Darnton, and framed a new way to answer it. In general, it's worth keeping this principle in mind.

But there was also a contingent, and problematic, reason for our success. Wealthy collectors and alumni had endowed the university where we worked with books, fellowships and research time for scholars. Access to Harvey's books, 20 years ago, was no easier to gain than it had been 50 years before. Readers needed credentials and permissions even to see them. Black and white microfilms were hard to read. And no article—even one as long as Lisa's and my treatment of Harvey—could give a full and vivid impression of the way in which Harvey's annotations circled around and climbed up the pages of his books, far less present their full, immense text, in the original and in translation, to readers around the world.

Lisa—always a strong advocate for underdogs—worried for years about these difficulties and limitations. One of her last major projects—undertaken in collaboration with Earle Havens of Johns Hopkins University and with the support of the Mellon Foundation—was the Archaeology of Reading. This collaborative enterprise—the historical part of which was based at the digital humanities unit that she founded in London, CELL (The Centre for Editing Lives and Letters) —has produced model collaborative editions, with digital facsimiles, transcriptions and

[12] Jardine and Grafton 1990.

[13] For the most expert of retrospective views and critiques, see Jardine 2016.

translations, of a substantial set of books annotated by Harvey and a slightly earlier Cambridge scholar, John Dee.[14] These include Harvey's Livy, which has already become the object of intensive study by other scholars.[15]

Over time, Lisa and I learned to see the utility of the history of reading in a second, very different light: as a continued reminder of the contingency and variety of intellectual history itself, and a preventative against—among other things—premature efforts at periodization, a practice of which both of us have at times been guilty. Even within the world of learned reading that continued to be the focus of our interests, reading practices varied extraordinarily, and reading experiences even more so. No one insisted more strongly than Lisa that many rabbit holes await future explorers of this territory—or would have been happier to see them revise her (and my) conclusions.

Lisa spent only a small part of her scholarly life in the margins. After her book on Erasmus appeared, she would rewrite the history of European, cultural, intellectual and scientific life in the sixteenth and seventeenth centuries. She continued to collaborate, not only with her team at CELL but also with Jerry Brotton, Alan Stewart and others. She showed how Hooke and Wren transformed the city of London and how the Huyghens family and their Dutch compatriots created a sparkling world of exotic gardens, spectacular works of art and penetrating inquiries into nature. And she told these stories in massive biographies and histories, as elegant and accessible in style as they were novel in content. Still, her work on marginalia, carried out with what now seem primitive scholarly tools, has been extraordinarily influential, even for her. More than half a century after she began work, more than a quarter century after our article on Harvey's Livy appeared, her work continues to shape research on books and readers in many areas of the humanities. Those margins turned out to be fertile, once Lisa taught us how to cultivate them.

Bibliography

Blair, Ann. 2010. *Too much to know: Managing scholarly information before the modern age*. New Haven: Yale University Press.

Bourland, Caroline Brown. 1940. Gabriel Harvey and the modern languages. *Huntington Library Quarterly* 4: 85–106.

Darnton, Robert. 1986. First steps towards a history of reading. *Australian Journal of French Studies* 23: 5–30.

Grafton, Anthony. 1979. Rhetoric, philology and Egyptomania in the 1570s: J.J. Scaliger's invective against M. Guilandinus's papyrus. *Journal of the Warburg and Courtauld Institutes* 42: 167–194.

———. 1983–1993. *Joseph Scaliger: A study in the history of classical scholarship*, 2 vols. Oxford: Clarendon Press.

[14] The Archaeology of Reading (https://archaeologyofreading.org/).

[15] Norbrook 2017.

Harvey, Gabriel. 1913. *Gabriel Harvey's marginalia*, ed. G.C. Moore Smith. Stratford-upon-Avon: Shakespeare Head Press.

Hirrel, Michael. 2012. When did Harvey write his famous note? *Huntington Library Quarterly* 75: 291–299.

Hutson, Lorna. 1989. *Thomas Nashe in context*. Oxford: Clarendon Press.

Jardine, Lisa. 1990. Inventing Rudolph Agricola: Cultural transmission, renaissance dialectic and the emerging humanities. In *The transmission of culture in early modern Europe*, ed. Anthony Grafton and Ann Blair, 39–86. Philadelphia: University of Pennsylvania Press.

———. 1993 (repr. 2015). *Erasmus, man of letters: The construction of charisma in print*. Princeton: Princeton University Press.

———. 2016. 'Studied for action' revisited. In *For the sake of learning: Essays in honor of Anthony Grafton*, ed. Ann Blair and Anja-Silvia Goeing, 999–1017. Leiden; Boston: Brill.

Jardine, Lisa, and Anthony Grafton. 1990. 'Studied for action': How Gabriel Harvey read his Livy. *Past & Present* 129: 30–78.

Luard, H.R., and Churchill Babington. 1864. *A catalogue of Adversaria and printed books containing MS. Notes preserved in the library of the University of Cambridge*. Cambridge: Cambridge University Press.

Norbrook, David. 2017. Rehearsing the plebeians: Shakespeare and the reading of Roman history. In *Shakespeare and the politics of commoners: Digesting the new social history*, ed. Chris Fitter, 180–216. Oxford: Oxford University Press.

Sherman, William. 1995. *John Dee: The politics of reading and writing in the English renaissance*. Amherst: University of Massachusetts Press.

Stern, Virginia F. 1972. The bibliotheca of Gabriel Harvey. *Renaissance Quarterly* 25: 1–62.

———. 1979. *Gabriel Harvey: His life, marginalia and library*. Oxford: Clarendon Press.

Wilson, Harold. 1948. Gabriel Harvey's method of annotating his books. *Harvard Library Bulletin* 2: 344–361.

Chapter 3
Dosing the Ailing Subject: Reconnecting Early Modern Health and Thought

Lisa Jardine

Abstract This chapter discusses early modern efforts to boost intellectual activity and control bodily functions. Starting with Robert Hooke's effort to conduct "trials" with chemical compounds, which he used as instruments for enhancing his mental capacities just as he used the microscope and telescope to enhance his vision, the chapter turns to Francis Bacon's personal purging routine, the details of which appear to have been suppressed by Bacon's later editors. Bacon's regimen of 'physic' aimed specifically at prolonging his life and entailed awareness of and a manifest interest in the effects of chemical medicine on the mind and body. Although not all medical men of the period were so enthusiastic for the educated gentleman's self-prescription of purgatives and toxic substances, Bacon and many others continued self-experimentation up to the time of their death. One of the many larger lessons to draw from these case studies linking health and thought is that interest in experimentation in the early modern period was, often, fundamentally concerned with the relation between mind and body. Indeed, there is a strong link joining early modern science and philosophy to experimental medical activity where theories of the human mind were being developed.

Keywords Robert Hooke · Francis Bacon · Self-experimentation · Prolongation of life · Theories of mind

> 6 February 1674: At Spanish coffee house tryd new mettall with Antimony, iron and lead. At Shortgraves. Tryd reflex microscope. –Robert Hooke[1]

The purpose of this conference is to reconnect health and thought in the early modern period. The two were not understood as separate in the seventeenth century;

[1] Hooke 1935, 85.

L. Jardine (✉)
University College London, London, UK
e-mail: gideon.manning@cgu.edu

G. Manning (ed.), *Testimonies: States of Mind and States of the Body in the Early Modern Period*, Archimedes 57,
https://doi.org/10.1007/978-3-030-39375-5_3

reuniting them will, I suggest, add a significant dimension to our historical understanding. I am interested in the way self-experiment, combined with careful attention to the states of mind which accompany states of body specifically altered by measured doses of chemically prepared substances generated a new and influential attitude to body/mind interactions.[2]

In the passage from Hooke's diary which provides my epigraph, what interests me, for a start, is the repetition of that word 'trial'. Hooke 'tried' a new toxic chemical compound as a medical remedy, and then 'tried' the reflex microscope. Both 'trials' involved skilled handling of the equipment/materials and careful observation of outcomes and accuracy. Two hands-on experiments in science and new technology. Both observed and recorded in the day-by-day laboratory notebooks of the habitual experimentalist, with himself as the experimental subject. Two domains within which Hooke expertly tests the latest thing in control and manipulation of the body and its senses. As Hooke makes clear in his preface to his handsome, illustrated scientific coffee-table book *Micrographia* (1665), he believed that new precision scientific instruments like the telescope and microscope would give back to mankind the clarity of perception lost at the Fall:

> It is the great prerogative of Mankind above other Creatures, that we are not only able to behold the works of Nature … but we have also the power of considering, comparing, altering, assisting, and improving them to various uses. And as this is the peculiar priviledge of humane Nature in general, so is it capable of being so far advanced by the helps of Art, and Experience, as to make some Men excel others in their Observations, and Deductions, almost as much as they do Beasts. By the addition of such artificial Instruments and methods, there may be, in some manner, a reparation made for the mischiefs, and imperfection, mankind has drawn upon it self.[3]

By that double use of 'trial' – Hooke's customary term for developing such experimentally-based 'helps' for the senses – we have at least a suggestion here that chemical 'instruments' might enhance man's post-lapsarian mental capacities, just as the microscope and telescope add magnification to his vision.[4]

---

[2] In spite of my saying that I am not interested in the seventeenth century as a drug culture as such, I am indebted, in the paper which follows, to the suggestive work of Sadie Plant, in her study of drug-linked sensibilities from the eighteenth century onwards (Plant 1999). I was also set thinking along the present lines by to two suggestive articles by Simon Schaffer broadly concerned with self-experiment: Schaffer 1992, 1998.

[3] Hooke 1665, preface.

[4] Among Hooke's posthumously published papers is one on the therapeutic use of 'bangue' or marijuana, particularly for insomnia. See Hooke 1726, 1–3: 'An Account of the Plant, call'd Bangue, before the Royal Society, Dec. 18. 1689. It is a certain Plant which grows very common in India, and the Vertues, or Quality thereof, are there very well known; and the Use thereof (tho' the Effects are very strange, and, at first hearing, frightful enough) is very general and frequent; and the Person, from whom I receiv'd it, hath made very many Trials of it, on himself, with very good Effect. 'Tis call'd, by the Moors, Gange; by the Chingalese, Comsa, and by the Portingals, Bangue. The Dose of it is almost as much as may fill a common Tobacco-Pipe, the Leaves and Seeds being dried first, and pretty finely powdered. This Powder being chewed and swallowed, or washed down, by a small Cup of Water, doth, in a short Time, quite take away the Memory and Understanding; so that the Patient undestands not, nor remembereth any Thing that he seeth,

My second starting-point is a passage written some 60 years before Hooke's, but equally concerned with state-of-mind-altering experiments. It comes from some jottings by Francis Bacon, which the editor of the definitive nineteenth century edition of Bacon's Works tucked away so that they would not attract notice.[5] In July 1608 Francis Bacon included in his (appropriately named) 'loose notes' or '*Commentarius solutus*' an account of the successive stages in one of his personal purging routines (elsewhere he explains that therapeutic medicines only work if they are incorporated in a programme which includes preparatory dosings and calming post-purging substances)[6]:

> After a maceration taken in the morning and working little I took a glister [enema] about 5 o'clock to draw it down better, which in the taking found my body full and being taken but temperate and kept half an hour wrought but slowly, neither did I find that lightness and cooling in my sides which many times I do, but soon after I found a symptom of melancholy such as long since with strangeness in beholding and darksomeness, offer to groan and sigh, whereupon finding a malign humour stirred I took three pills of aggregative corrected according to my last description, which wrought within two hours without griping or vomit and brought much of the humour sulphurous and fetid, then though my medicine was not fully settled I made a light supper without wine, and found myself light and at peace after it. I took a little of my troch[ises] of [sal] amon[iac] after supper and I took broth immediately after my pill.[7]

---

heareth, or doth, in that Extasie, but becomes, as it were, a mere Natural, being unable to speak a Word of Sense; yet is he very merry, and laughs, and sings, and speaks Words without any Coherence, not knowing what he saith or doth; yet is he not giddy, or drunk, but walks and dances, and sheweth many odd Tricks; after a little Time he falls asleep, and sleepeth very soundly and quietly; and when he wakes, he finds himself mightily refresh'd, and exceeding hungry. And that which troubled his Stomach, or Head, before he took it, is perfectly carried off without leaving any ill Symptom, as Giddiness, Pain in the Head or Stomach, or Defect of Memory of any Thing (besides of what happened) during the Time of its Operation. ... The Plant is so like to Hemp in all its Parts, both Seed, Leaves, Stalk, and Flower, that it may be said to be only Indian Hemp.... [T] his I have here produced, is so well known and experimented by Thousands; and the Person that brought it has so often experimented it himself, that there is no Cause of Fear, tho' possibly there may be of Laughter. It may therefore, if it can be here produced, possibly prove as considerable a Medicine in Drugs, as any that is brought from the Indies; and may possibly be of considerable Use for Lunaticks, or for other Distempers of the Head and Stomach, for that it seemeth to put a Man into a Dream, or make him asleep, whilst yet he seems to be awake, but at last ends in a profound Sleep, which rectifies all; whereas Lunaticks are much in the same Estate, but cannot obtain that, which should, and in all Probability would, cure them, and that is a profound and quiet Sleep.'

[5]On Spedding's commitment in his edition of the *Works of Francis Bacon* to rescuing Bacon's reputation from Macaulay's, in Spedding's view, libellous slurs, see Jardine and Stewart 2000. A small collection of 'medical remains', also comprising Bacon's personal remedies – this time labelled 'Grains of Youth' – is equivalently obscurely placed at the very end of the *Philosophical Works* (Spedding et al. 1857–1874, 3: 827–36). Among the regular activities Bacon lists here 'for his own use, out of the book of the prolongation of health' are the following: 'Once in the week, or at least in the fortnight, to take the water of mithridate distilled, with three parts to one, or strawberry-water to allay it; and some grains of nitre and saffron, in the morning between sleeps. ... A cooling clyster to be used once a month, after the working of the maceration [for which the recipe is attached] is settled. ... To use once during supper time wine in which gold is quenched'.

[6]*Sylva sylvarum*, Century I, 65; Spedding et al. 1857–1874, 2: 368.

[7]See Jardine and Stewart 1998, 300–301. 'Pills' in the period are always purgatives (trochises are a type of pill). Tucked away elsewhere in Bacon's paper we find recipes for remedies like the ones

In the margin of the manuscript, Bacon noted, in partial explanation of his experience: 'Note, there had been extreme heats for ten days before and I had taken little or no physic' – external factors, too, may contribute to the state of mind he achieves at the end of his 'regimen'.

Here, in the informality (or unguardedness) of his private jottings, Bacon records his therapeutic purging 'without any matter of restraint' (as he put it in the introduction to this manuscript). So he includes the moods induced at successive stages in his self-dosing: his 'glister' produces 'a symptom of melancholy such as long since with strangeness in beholding and darksomeness, offer to groan and sigh'; the light supper without wine which he takes after purging makes him feel 'light and at peace'. Something is going on here beyond the administering of therapeutic medicaments to treat real or imagined ailments, I suggest. I want to propose that those committed to the regimen of 'physic' – the same band of people who pursue the new mechanistic, technologically led science of the later Royal Society – were as aware of, and interested in, the altering effects of chemical medicine on the mind as they were in its effects on the body.

And I add here a further suggestive quote from Hooke's diary, to complete the array of mind/body mutual activity to which I will be giving attention. This one (entered on 31 May 1674) indicates a perceived connection between the effects on

---

he describes himself taking here. 'A cooling clyster to be used once a month, after the working of the maceration is settled. Take of barley-water, in which the roots of bugloss are boiled, three ounces, with two drams of red sanders [sandalwood], and two ounces of raisins of the sun, and one ounce of dactyles, and an ounce and a half of fat caricks; and let it be strained, and add to it an ounce and a half of syrup of violets: let a clyster be made. Let this be taken <with veal> in the aforesaid decoction' (Spedding et al. 1857–1874, 3: 833). For the maceration we can refer to Rawley's *Life of Bacon*: 'You may be sure that he would not neglect that himself, which he so much extolled in his writings, and that was the use of nitre; whereof he took in the quantity of about three grains in thin warm broth every morning, for 30 years together next before his death. And for physic, he did indeed live physically, but not miserably; for he took only a maceration of [six drachms of] rhubarb, infused into a draught of white wine and beer mingled together for the space of half an hour, once in 6 or 7 days, immediately before his meal (whether dinner or supper), that it might dry the body less; which (as he said) did carry away frequently the grosser humours of the body, and not diminish or carry away any of the spirits, as sweating doth' (ibid., 1: 17). Elsewhere Bacon himself notes variations on this 'maceration': 'To add to the maceration six grains of cremor tartari, and as much enula [elecampane root]' (a further note in Spedding's version of this text adds a variant on the quantities: three grains of tartar and two of enula 'to cut the more heavy and viscous humours; lest rhubarb work only upon the lightest') (ibid., 3: 833). In the Lambeth Palace Library Manuscripts we find the following among Bacon's last secretary, William Rawley's reminiscences: 'In my Lords Maceration, the Quantities are; of Rhubarb a dram and half; of Roman wormwood, and Squinanth, each 20' (Lambeth Palace Library ms 2086, p. 38). In a further variant on the maceration Bacon himself proposes in some notes on preserving his own health 'to take once in the month at least, and for 2 days together, a grain and a half of castor in my broth, and [at?] breakfast' (Spedding et al. 1857–1874, 3: 833). A final part of his daily routine was (according to this schedule) 'To take every morning the fume of lign-aloes, rosemary and bays dried, which I use; but once in a week to add a little tobacco, without otherwise taking it in a pipe' (ibid.). The note 'to appoint every day an hour *ad affectus intentionales et sanos. Qu. de particulari*' presumably complies with his view that sexual excitement – but not ejaculation – is life-prolonging.

the body of certain therapeutic substances and an increased clarity of the intellectual faculties:

> Took Childs vomit, Infusion Croc[us] Met[allicus]. ioz. Wrought pretty well. Refresht by it. D[ined] H[ome]. Made description of quadrant. Lost labour. Slept after dinner. A strange mist before my eye. Not abroad all day. Slept little at night. My fantcy very cleer. Meditated about clepsydra, quadrant, scotoscopes, &c.[8]

A dose of a proprietary emetic and a mildly toxic purgative produce clouded vision, wakefulness, and clarity of 'fancy' – thought or imagination. As a result, Hooke has particularly vivid and interesting thoughts about water-clocks, navigational instruments and tools for microscopy (his diary shows him repeating 'physic' which has thus enhanced his mental faculties).

The context for both my opening passages is the 'new' chemical medicine, whose popularity, in the early seventeenth century, more traditional physicians deplored. In 1644, when the young John Aubrey was considering going to Italy to study medicine, he approached the elderly William Harvey – discoverer of the circulation of the blood, and royal physician the Charles I – for advice. What preparatory reading should he do? Harvey's response was characteristically to-the-point: 'he bid me goe to the fountain head, and read Aristotle, Cicero, Avicenna, and did call the neoteriques shitt-breeches'.[9] Aubrey also tells us that Harvey 'did not care for chymistrey, and was wont to speak against them with an undervalue'.[10]

These 'shitt-breeches' were the fashionable prescribers patronised by persons-about-town and the intellectual virtuosi, medical men who embraced the 'new' treatments and remedies which tend to be bundled together under the catch-all label 'Paracelsian'.[11] Their pursuit of a new medicine or 'physic' was coloured by their

---

[8] Hooke 1935, 105.

[9] Cit. Siraisi 1987, 353. For another colourful text connecting 'shit' and Paracelsianism, see Parfitt 1975, Epigram cxxxiii: 'On the famous voyage', which tells the story of an ill-advised journey by wherry along the effluent-filled Thames at night: 'At this a loud/ Crack did report itself, as if a cloud / Had burst with storm, and down fell, *ab excelsis*, / Poor Mercury, crying out on Paracelsus, / And all his followers, that had so abused him: / And, in so shitten sort, so long had used him: / For (where he was the god of eloquence, / And subtlety of metals) they dispense / His spirits, now, in pills, and eek in potions, / Suppositories, cataplasms and lotions.' I am grateful to Andy Gordon for this reference.

[10] Cit. Clark 1964, 1: 302.

[11] On this tendency for Paracelsus to be used as a catch-all designation, rather than as a reference to a particular 'school' of medical theory see Multhauf 1954, 125–6: 'Medical chemistry, which, in the middle of the sixteenth century, was still a matter of ill-defined distillates and occult specifics, seems to have come of age during the half century thereafter, for in Croll's *Basilica chymica* and other contemporary works we encounter many of the chemical medicines which were to play so great a role in the medicine of the next two centuries. Four mineral materials, vitriol, mercury, antimony, and gold, dominate the pharmacopoeia of Croll, but it was not the utilization of these minerals that distinguished his book, for they appear to have been the most highly regarded materials of the medical chemist as early as the middle fourteenth century. It was rather the face that in the work of Croll and in those of his contemporaries we find for the first time a real *chemical* procedure in the production of medicines derived from these minerals. This procedure consisted in general in their dissolution in the several mineral acids, followed by recovery of the inorganic salts

involvement with other 'new' scientific activity (chemistry, microscopy, mechanistic philosophy) – the shared interests of these men (and a few women) led at the Restoration to the establishing of the Royal Society in London.[12]

Ironically, in spite of Harvey's disapproval, the efficacy of the 'neoteriques''s prescriptions depended in no small measure upon their embracing and applying Harvey's groundbreaking medical 'discovery' of the circulation of the blood. Five or so years before Harvey publicly announced, in his *De motu cordis* (1628), the theory that the blood circulates,[13] Bacon already clearly had some kind of circulatory 'flow' or bloodstream in mind when he wrote in his 'History of Life and Death':

> I fully believe that if something could be infused in very small portions into the whole substance of blood, over which the action of the spirit and heat should have little or no power [he is thinking of 'gold-leaf or filings, or powder of pearls, gems and coral, or the like'], it would stop not only all putrefaction, but arefaction [drying] likewise, and be very effectual in prolonging life.[14]

Bacon proposes giving such remedies by mouth, whence the therapeutic substance would (he hoped) enter and suffuse the blood. Thirty or so years later the Oxford physicians in Robert Boyle's circle would be administering opiates and toxic metal salts intravenously, taking it entirely for granted that the substance inserted (via a slender silver, sharp-tipped funnel and a bladder) would swiftly be carried by the blood around the entire body, rapidly taking its narcotic or purgative effect.[15] In his illustrated treatise *Micrographia* (1665), Hooke showed that the so-called 'gravel' in urine was actually tiny crystals, which he found to be soluble in oil of vitriol (sulphuric acid). He suggested that this discovery might offer a remedy for that most wide-spread of seventeenth-century complaints, 'the stone'. What was required was a solvent for the crystals harboured by the blood, which could be injected directly:

---

prepared. The definition of the series of chemical products yielded by these procedures, and their successful introduction into the pharmacopoeias of Europe, was accomplished by the physician-chemists known as "the Paracelsians." The works of these men show that the appellation was no misnomer, for with few if any exceptions they exhibit an enthusiastic adherence to the doctrines of Paracelsus. But their association with the remedies of Paracelsus is by no means so clear, despite the existence of / evidence that it was his alleged cures which first attracted their interest. … Moreover, although the material of most common interest, vitriol, was chiefly used by both Paracelsus and the Paracelsians in the acid form of "spirit of vitriol", Paracelsus usually used it as a medicine per se while his followers usually used it as a reactant for the preparation of mineral remedies.' See also Debus 1965, 137: 'Few doctors of repute in England seriously considered Paracelsian speculative cosmology, but the benefits of chemical medicines were utilized by all.'

[12] Previously, through the Commonwealth and Protectorate period, these interests were represented within the Royal Society of Physicians – retitled the Society of Physicians. Women were prominent in therapeutic practice beyond the formal medical profession, during and after the civil war period, but were excluded from the medical colleges and guilds. See Hunter and Hutton 1997.

[13] Harvey first outlined the theory of blood circulation in the Lumleian Lecture to the Royal College of Physicians in 1616.

[14] *History of Life and Death*; Spedding et al. 1857–1874, 5: 289.

[15] Christopher Wren's reputation as a skilled medical practitioner in Oxford in the 1650s was based on his techniques for intravenous injection (Jardine 2002, 122–4).

> How great an advantage it would be to such as are troubled with the Stone, to find some menstruum that might dissolve them without hurting the Bladder, is easily imagin'd, since some injections made of such bodies might likewise dissolve the stone, which seems much of the same nature.[16]

One whole section of the first Century of Bacon's *Sylva sylvarum* is devoted to 'experiments in consort touching purging medicines', in the new medical tradition Harvey so bluntly despised. Bacon, too, is brusquely matter-of-fact (not for them any Victorian mincing of words where bodily functions are concerned):

> There be divers medicines which in greater quantity move stool, and in smaller urine: and so contrariwise, some that in greater quantity move urine, and in smaller stool. Of the former sort is rhubarb [root], and some others. The cause is, for that rhubarb is a medicine which the stomach in a small quantity doth digest and overcome (being not flatuous nor loathsome) and so sendeth it to the mesentery veins[17]; and so being opening, it helpeth down urine: but in a greater quantity, the stomach cannot overcome it, and so it goeth to the guts. Pepper by some of the ancients is noted to be of the second sort; which being in small quantity, moveth wind in the stomach or guts, and so expelleth by stool; but being in greater quantity, dissipateth the wind; and itself getteth to the mesentery veins, and so to the liver and reins; where, by heating and opening, it sendeth down urine more plentifully.[18]

Characteristically, the tone and tenor of this discussion signals that Bacon is basing his observations on direct personal experience[19]; characteristically, too, the process of medicaments is described in mechanistic sequence, as successive stages of cause and effect.

Francis Bacon and his brother Anthony had both suffered fragile health from boyhood, and were early habituated to medicinal dosings of one kind and another (too much physic-taking eventually ruined Antony's career prospects).[20] Passing references in surviving correspondence reveal that both were committed to a routine of regularly dosing themselves with therapeutic remedies, in a 'modern' pattern of combined cathartics and stimulants. A letter from Anthony Bacon to his mother of 1593 typically alludes to the customary use of 'physic-taking' – routine clearing out

---

[16] On early intravenous injection and blood transfusion see Gunther 1926, 106–7: 'Wren first suggested to Boyle at Oxford not later than 1656 to ligature the veins of a living animal, open them on the side of the ligature nearest the heart, and inject with "slender syringes or quills fastened to bladders". He recommended as the subject "pretty big and lean dogs". This operation appears to have been frequently practised in Oxford, and also in London before the Royal Society. "And they hope likewise, that beside the *medical* uses, that may be made of this *invention*, it may also serve for *anatomical* purposes, by filling, after this way, the vessels of an animal as full as they can hold, and by exceedingly distending them, discover new vessels." … Wren, then a youth of twenty-four, injected wine and ale into the blood of a living dog by one of the veins, and noted that the animal became extremely drunk. The experiment he takes "to be of great concernment and what will give great light to the theory and practice of physic"'.

[17] The veins in the wall of the small intestine.

[18] Spedding et al. 1857–1874, 3: 358.

[19] For the varieties of locution used by Bacon in his 'histories' to indicate that the evidence is derived from his own experimental observations, or is aduced from that of others, see Jardine and Silverthorne 2000.

[20] See Jardine and Stewart 1998.

of the body using toxic substances to stimulate vomiting and excreting. First he pleads for his mother to help his brother Francis out of a tight corner by clearing his debts (both Bacons were hopelessly bad with money), and then he turns to his personal state of health:

> Touching myself my diet, I thank God hitherto, hath wrought good effect, and am advised to continue this month not meddling with any purgative physic more than I must needs, which will be but thrice during my whole diet.[21]

It is hardly surprising that Francis Bacon dabbled in self-administered remedies, self-dosing was something of a vogue for those with the means to acquire exotic imported pharmaceuticals. It is more surprising, perhaps, that he developed this into a significant part of his preliminary excursions into the 'new sciences' (and that this part of his scientific oeuvre has gone largely unremarked). His *Sylva sylvarum* (his first attempt at a compendious natural history on which to base a methodical approach to knowledge of nature) is liberally sprinkled with observations relating to a wide range of self-administered medicaments, as is the fragment of a *History of Life and Death* which was one of the papers he was working on at the time of his own demise.

In the *History of Life and Death*, opiates and stimulants are prominent among the substances recommended for 'the operation upon the spirits, that they may retain their youth and renew their vigour'.[22] For example:

> Opium is by far the most powerful and effectual means of condensing the spirits by flight [and thereby prolonging the youth of a man]; and next to it opiates and soporifics in general.
>
> The power of opium to condense the spirits is very remarkable; for perhaps three grains will in a short time so coagulate them that they cannot separate, but are quenched and rendered immoveable. …
>
> The flight of the spirits by means of opium and opiates is best seen when they are applied externally; for the spirits instantly retire and will return no more, but the part mortifies and turns to a gangrene.
>
> Opiates give relief in great pain, as the stone, or amputation of a limb; principally by putting the spirits to flight.[23]

It is for these powers to 'condense the spirits' and aid longevity that Bacon recommends both coffee and tobacco, rather than as stimulants[24]:

---

[21] Lambeth Palace Library ms 649, 103–04. AB Anthony Bacon to Lady Bacon, 16 April 1593, Gray's Inn.

[22] Spedding et al. 1857–1874, 5: 268–83.

[23] Ibid., 5: 270.

[24] 'Opium and opiates are clearly found to excite the sexual passion, which shows their power to strengthen the spirits. … For this is common to opiates, as the spirits being strengthened and condensed will fight against any disease. The Turks use likewise a kind of jerb called "coffee," which they dry, grind to powder, and drink in warm water. They affirm that it gives not small vigour both to their courage and their wit. Yet this taken in large quantities will excite and disturb the mind; which shows it to be of a similar nature to opiates' (ibid., 5: 270–1).

> The use of tobacco has immensely increased in our time. It affects men with a kind of secret pleasure, so that persons once accustomed to it can scarce leave it off. It tends no doubt to relieve the body, and remove weariness; and its virtue is commonly thought to lie in this, that it opens the passages and draws off the humours. But it may be more properly referred to the condensation of the spirits; for it is a kind of henbane, and manifestly affects the head, as all opiates do.
>
> Humours are sometimes generated in the body, which are a kind of opiates themselves; as is found in some kinds of melancholy, wherewith if a man be seized, he is very long-lived.[25]

And again:

> Remembering what I have said before concerning opiates, nitre, and the like, which so powerfully condense the spirits, I do not think it would be amiss if once a fortnight three or four grains of castor be taken in a morning broth, with a little angelica seed and calamus aromaticus. For these both strengthen / the brain, and excite in that density of the substance of the spirits (which is so necessary for longevity) vivacity and vigour of motion.[26]

So Francis Bacon writes from first-hand experience of both physic-administering, and its consequences for his own bodily system, when he details the merits and effectiveness of various types of purgatives amongst the medical materials gathered for short-term scientific use in his scientific compendium, the *Sylva sylvarum*.[27] He emphatically rejects sympathy-based explanations of the way cathartics act on the body:

> The operation of purging medicines, and the causes thereof, have been thought to be a great secret; and so according to the slothful manner of men, it is referred to a hidden propriety, a specifical virtue, and a fourth quality, and the like shifts of ignorance.

On the contrary, he writes, the way purgatives act can be accounted for by the direct action of the substance on the stomach; the type of action depending on the type of medicine:

> The causes of purging are divers: all plain and perspicuous, and thoroughly maintained by experience. The first is, that whatsoever cannot be overcome and digested by the stomach, is by the stomach either put up by vomit, or put down to the guts; and by that motion of expulsion in the stomach and guts, other parts of the body (as the orifices of the veins, and the like) are moved to expel by consent. For nothing is more frequent than motion of consent in the body of man. This surcharge of the stomach is caused either by the quality of the medicine, or by the quantity. The qualities are three: extremely bitter, as in aloes, coloquintida, &c.; loathsome and of horrible taste, as in agaric, black hellebore, &c.; and of secret malignity and disagreement towards man's body, many times not appearing much in the taste; as in scammony, mechoacham, antimony, &c. And note well, that if there by any medicine that purgeth, and hath neither of the first two manifest qualities, it is to be held

[25] Ibid., 5: 271.

[26] Ibid., 5: 299–300.

[27] I stress this 'hands-on' quality of what follows, because some scholars (e.g., Findlen 1997) have suggested that Bacon's *Sylva sylvarum* is largely made up of 'experiments' drawn from a restricted body of 'natural history' writing. As I have explained elsewhere (Jardine and Silverthorne 2000), Bacon is meticulous in indicating when he has conducted experiments himself – his medical observations are a clear case in point.

> suspected as a kind of poison; for that it worketh either by corrosion, or by a secret malignity and enmity to nature; and therefore such medicines are warily to be prepared and used.[28]

The attempt here at classifying *materia medica* as cause and effect is clear. Nevertheless, in spite of this attempted 'modern' account of the workings of cathartic substances, at the point at which Bacon asks the reader to 'note well' that some things that purge are true poisons, his explanation regresses to precisely the kind of 'slothful manner' he explicitly disdained at the beginning of the passage – falling back on a vocabulary of 'secret malignity and enmity to nature', to explain what renders these dangerous.[29]

Undeterred by such occasional lapses, the next generation of natural philosophers ('scientists' in modern-day parlance) studied medical entries like this one in Bacon's *Sylva sylvarum* closely.[30] The Oxford physician John Ward took careful notes on this very passage in his medical diary. There we find him (writing in the 1640s) quoting verbatim from the passage I just read:

> Causes of purgatives are represented as (1) Extreme bitter as in Aloes and Colloquinta. (2) Loathsome and horrible taste as Agarick and black hellebore. (3) By secret malignity many times not appearing in the taste as Scammony and Antimony: and if anything purge which hath not one of these 2 former virtues in it, it is to be suspected for poison. If we drink a great quantity of new milk it purgeth; that a mordication or vellitation of the orifice of the veins especially of the Mesentary veins: that almost all purges cause a kind of twitching etc.[31]

Ward treats the passage from Bacon's *Sylva sylvarum* as if it were indeed a compilation of important materials for a specifically seventeenth century 'modern' medical practice.[32] The job of the physician, on this model, is to apply the ancient *materia medica* (with significant 'new' chemically extracted additions) experimentally, modifying and combining the old resources so as to extend the repertory of available cures by reasoned transpositions, combinations and comparisons. This represents a 'modern' attitude to remedies, producing applications not attempted by the authors of the ancient compilations.

---

[28] Spedding et al. 1857–1874, 2: 355–6.

[29] My own view is that it is this kind of verbal slippage which leads to the secondary literature's including the kinds of passage I am dealing with here as part of their treatment of seventeenth century 'occult' rather than sense-related qualities. See, for example, Gouk 1999. Hutchison 1997 redefines 'occult' as 'insensible', but nevertheless excludes consideration of such qualities from his discussion of 'science'.

[30] Penny Gouk draws attention to a copy of the *Sylva sylvarum* copiously annotated by Dean Wren (Christopher Wren's father) in the 1640s (Gouk 1999). Hooke based a number of the experiments he performed as curator before the members of the Royal Society directly on *Sylva sylvarum* experiments.

[31] Cit. Gunther 1926, 45–46. Gunther erroneously records these as being remarks made in a lecture attended by Ward, given by the Regius Professor at Oxford Thomas Clayton.

[32] Ward's diaries as a whole are a series of practical workbooks, compilations of data based on observing the work of Boyle, Wren and the other virtuosi he met, as he embarked on his training for medical practice himself. See Frank 1974.

The physician, on this model, takes the *materia medica* (medical materials) from the traditional classical sources, adds a significant body of 'new' chemically extracted substances, and uses them experimentally, modifying and combining the old resources so as to extend the repertory of available cures by reasoned transpositions, combinations and comparisons. Habitually, he tries them out on himself, since this is the most reliable way of gaining a clear idea of the effect of the remedy as experienced by the patient.

As Harvey's 'shitt-breeches' comment made clear, not all medical men in the period shared the enthusiasm of educated gentlemen for self-prescription with purgatives and metal- or chemical-based toxic substances. There were those around Bacon who were concerned at his habit of self-prescribing and excessive physic-taking. His nephew Edmund Bacon – a trained medical man who calls himself Francis's 'faithful apothecary' and who owned a 'laboratory' – wrote anxiously to him on 19 October 1616 (this letter has once again been tucked away out of sight by editors of Bacon)[33]:

> My Lord, I am bold to present into your hands by this bearer, whom the Law calls up, some salt of wormwood [absinthe], being uncertain, whether the regard of your Health makes you still continue with the use of that Medicine. I would wish it otherwise; for I am perswaded, that all Diuretics, which carry with them that punctious nature & caustic quality by calcination are hurtfull to the kidneys, if not enemies to other principal parts of the body.[34]

There is no evidence of Bacon's taking any notice of such advice – as we shall see, he continued with regular doses of 'caustic' medicaments right up to the time of his death.[35]

Like Ward, Hooke drew on Bacon's *Sylva sylvarum* for some of his own practical experiments.[36] Where physic was concerned, however, Hooke's experiments were performed outside the Society, in the circle of like-minded physic-taking friends he frequented. In his diary he noted meticulously the effects of *materia medica* on sick

---

[33] I am implying (as Alan Stewart and I have elsewhere in Jardine and Stewart 1998) that Bacon's editors suppressed evidence of his practical involvement with scientific experiment, particularly his dabblings in metallurgy, alchemy and medicine.

[34] The letter continues: 'Wherein, if it shall please you, for your better satisfaction, to call the advice of your learned physicians, & that they shall resolve of any medicine for your health, wherein my poor Labour may avail you, you know where your faithfull Apothecary dwells, who will be ready at your Commandment, as I am bound hath by your favours to myself as also by those to my Nephew, whom you hath brought out of darkness into light, & by what I hear, have already made him by your bounty a subject of emulation to his elder brother. We are all partakers of this your kindness toward him & for myself, I shall be very ready to deserve it by any service that shall lie in the power of your Ldp.' I am extremely grateful to Dr. Patricia Brewerton for drawing my attention to this letter, discovered in Birch's manuscript of documents he had prepared for the printer, but not subsequently included in his edition (BL ms Additional 4262 folio 57r-v).

[35] On which death, as it relates closely to the present argument, see Jardine and Stewart 1998, 502–11.

[36] In July 1680, Hooke performed an experiment based on §9 at a meeting of the Society (cit. Gouk 1999, 160).

friends; above all he recorded symptom, remedy and outcome in experiments in which the patient was also the experimenter – self-experiment.

In November 1672, Robert Hooke recorded in his diary that one of his most respected friends, the Royal Society's founding figure Bishop John Wilkins, was gravely ill 'of the stone', as it was thought.[37] Three days later Hooke reports: 'Lord Chester [Wilkins] desperately ill of the stone, stoppage of urine 6 dayes. oyster shells 4 red hot quenched in cyder a quart and drank, advised by Glanvill. Another prescribed *flegma acidum succini rectificatum* cum sale tartari [a phlegm-releasing compound of acid of amber rectified with tartaric salt]. Dr. Godderd advisd Blisters of cantharides [Spanish fly] applyd to the neck and feet or to the vains.'.[38] Two days later Hooke was once more at the Bishop's bedside, Wilkins was 'desperately ill and his suppression continued'; the following day Wilkins died 'about 9 in the morning of a supression of urine'. An autopsy was conducted the next morning: 'Dr Needham brought in account of Lord Chesters having no stoppage in his uriters nor defect in his kidneys. There was only found 2 small stones in one kidney and some little gravell in one uriter but neither big enough to stop the water. Twas believed his opiates and some other medicines killed him, there being no visible cause of his death, he died very quickly and with little pain, lament of all'.[39]

Hooke and his medical colleagues administer highly toxic 'physic' to the body of the sick man, and watched for alleviating symptoms which do not come. Chemically, these acidic remedies can be thought of as combining in the blood to dissolve the chalky solids of the stone. One of the phenomena whose greatly magnified image Hooke inspected, and reproduced in his popular illustrated treatise on the microscope, *Micrographia* (1665), was his own clouded urine – urine which contained, in suspension, a chalky substance, or 'gravel'. Hooke announced that close examination revealed this 'gravel' to be crystalline:

> I have often observ'd the Sand or Gravel of Urine, which seems to be a tartareous substance, generated out of a Saline and a terrestrial substance crystalliz'd together, in the form of Tartar, sometimes sticking to the sydes of the Urinal, but for the most part sinking to the bottom, and there lying in the form of coorse common Sand; these, through the Microscope, appear to be a company of small bodies, partly transparent, and partly opacous, some White, some Yellow, some Red, others of more brown and duskie colours.

Hooke found that the crystals could be made to dissolve again in 'Oyl of Vitriol [concentrated sulphuric acid], Spirit of Urine [impure aqueous ammonia, containing ammonium carbonate], and several other Saline menstruums'. Might not this offer a remedy for that most wide-spread of seventeenth-century complaints, 'the stone', he suggested? What was required was a solvent for the crystals harboured by the blood, which could be injected directly into the bladder:

---

[37] Hooke 1935, 13.

[38] Ibid.

[39] Hooke 1935, 14. It was almost another 200 years before the medical profession came to understand the significance of the structure of the crystalline deposits in gout, and developed a genuine cure (See Porter and Rousseau 1998, chapter 5).

> How great an advantage it would be to such as are troubled with the Stone, to find some menstruum that might dissolve them without hurting the Bladder, is easily imagin'd, since some injections made of such bodies might likewise dissolve the stone, which seems much of the same nature.[40]

In *Micrographia*, Hooke went on to suggest that Wren's method for injecting liquids into the veins of animals (which Hooke hinted had been more widely tried than had been reported, by Boyle among others) might be adapted to introduce solvents into the bloodstream of those suffering from a variety of common ailments:

> Certainly, if this Principle were well consider'd, there might, besides the further improving of Bathing and Syringing into the veins, be thought on several ways, whereby several obstinate distempers of a humane body, such as the Gout, Dropsie, Stone, &c. might be master'd and expell'd.[41]

Failure of the treatment in Wilkins's case is explained, for Hooke, by the fact that the diagnosis on Wilkins's inability to pass urine was incorrect: there was no stoppage in his urethra. Far from curing the condition, as Hooke laconically observes, it was the powerful toxins which probably proved fatal. The medical 'experiment' had failed, lending some irony to the report that as Wilkins lay dying, 'he said he was "ready for the Great Experiment"'.[42]

Similar attempts at observing dose and outcome in others, and relating them as cause and effect are recorded elsewhere in Hooke's diary:

> 4 July 1673: 'This evening Sir R. Moray died suddenly being choked with flegme in indeavouring to vomit. he had Dind at Lord Chancellers and about an howr before his Death drank 2 glasses of cold water.'
>
> 30 October 1673: 'at Dr. Wrens he very sick with physick taking day before.'[43]

Most of Hooke's diary entries concerning physic-taking, however, have to do with minutely recording his own consumption and outcomes. Here are a few typical examples:

> 27 April 1674: Took Dr. Thomsons vomit [later: Slayer told me that Dr. Thomsons vomit was Sneider's *Antimonia Tartarizatis*]. It vomited twice. Purged 10 or 12 times. Read over discourse against Hevelius. At Garaways till 10. Made me sleep ill, and made my armes paralytick with a great noyse in my head. I had also some knawing at the bottom of my belly.[44]
>
> 30 July 1675: Took SSA [spirit of sal ammoniac] with small beer at Supper. Very feavorish all night but slept well. Strangely refresht in the morning by drinking SSA with small beer and sleeping after it, it purged me twice. [and the next morning] In a new world with new medicine.
>
> 1 August 1675: Took volatile Spirit of Wormwood which made me very sick and disturbed me all the night and purged me in the morning. Drank small beer and spirit of Salamoniack. I purged 5 or 6 times very easily upon Sunday morning. This is certainly a great

[40] Hooke 1665, 81–2.

[41] Ibid., 144–5.

[42] Crowther 1982, 47.

[43] Hooke 1935, 49, 67.

[44] Ibid., 99.

> Discovery in Physick. I hope that this will dissolve that viscous slime that hath soe much tormented me in my stomack and gutts. *Deus Prosperat*.[45]

Shortly after this, however, sal ammoniac ceased to produce satisfactory therapeutic results for Hooke, and he moved on to poison his system with new toxic remedies.

As we eavesdrop on Hooke's self-experimenting with toxic purgatives, the patterns of dosing himself closely resemble his (and Bacon's) experiments in other, more conventional scientific domains. Substances used are administered with close attention to causal relations between measured dose, combinations of substances, and direct bodily effects. Accompanying mental states – 'strangely refresht', 'Refresht by it', 'much disturbed my head', 'made me cheerful' – are carefully noted. We recall again Bishop John Wilkins's supposed deathbed comment that he was 'ready for the Great Experiment', following the failed lesser experiments of administering solvents to disperse his 'stone', and that he was in 'little pain'.

It is also worth noting that a regimen of systematic self-dosing with purgatives was habit-forming, not least because vomiting or excreting were commonly preceded by a state of elation (associated with the chemical side-effects of the drug).[46] On 16th February, 1673, Hooke took Andrews cordial, which purged him:

> [It] brought much slime out of the guts and made me cheerful. Eat dinner with good stomach and pannado at night but drinking posset upon it put me into a feverish sweat which made me sleep very unquiet and much disturbed my head and stomach. Taking sneezing tobacco about 3 in the morn clear my head much and made me cheerful afterwards I slept about 2 hours, but my head was disturbed when I waked.[47]

Only very rarely, as, for instance, on the 3rd of August, 1673, a Sunday, does Hooke record that he 'took no physic'.[48] On the basis of the evidence in Hooke's diary, one might, I think, suggest that here we have a regimen (a meticulously planned and observed programme of food and drugs intake, over extended periods) which serves two functions simultaneously: it regulates bodily functions (meanwhile producing side-effects like numbness and blurred vision which are then treated with further ingestion of substances); and it excites the mental faculties, producing a clarity

[45] Ibid., 172. Entries of this kind can be matched from Anthony Bacon's, Francis Bacon's and Boyle's letters, and from Ward's, Pepys's and Aubrey's diaries.

[46] A number of the remedies used by Bacon, Hooke and others are still listed among modern alternative medical therapies as 'combating melancholy' when administered in small doses. Wormwood (absinthe) is just one of the remedies which elates when imbibed, but over time acts on the nervous system, producing tremors and delusions.

[47] Sometimes the remedy's intoxicating properties – its ability to give the patient a 'high' – were discovered accidentally. Hooke's friend and colleague Edmond Halley was so taken with the clear-headedness and calming effect of an over-large dose of opium he had inadvertently ingested (because it had been insufficiently mixed into a compound he had taken) that he read a paper on it to the Royal Society in January 1690, describing how 'instead of sleep, which he did design to procure by it, he lay waking all night, not as if disquiet with any thoughts but in a state of indolence, and perfectly at ease, in whatsoever posture he lay' (Halley 1937, 217).

[48] Cit. Beier 1989, 244.

which is conducive to slightly fevered intellectual activity of the kind for which Hooke had something of a reputation.

Armed with this kind of knowledge of Bacon's long-term experimental programme in purgative self-dosings, we can look with a fresh eye at the sections of Bacon's *Sylva sylvarum* devoted to 'animal spirits' – a favourite topic with philosophers of mind in the period, but one which we tend to associate with generalised early modern versions of 'action at a distance' of all kinds, from sensation to religious inspiration. Bacon notes frankly that contemporary accounts of the workings of 'animal spirits' are woefully inadequate. They are 'like prospectives [perspective paintings] that shew things inward, when they are but paintings'. Animal spirits, Bacon insists, are real, material substances, not artistic imaginings or metaphors – 'Neither is this a question of words, but infinitely material in nature':

> For spirits are nothing else but a natural body, rarified to a proportion, and included in the tangible parts of bodies, as in an integument. ... and they are never (almost) at rest; and from them and their motions principally proceed arefaction, colliquation, concoction, maturation, putrefaction, vivification, and most of the effects of nature.
>
> ... As for the differences of tangible parts in bodies, the industry of the chemists hath given some light, in discerning by their separations the oily, crude, pure, impure, fine, gross parts of bodies, and the like. And the physicians are content to acknowledge that herbs and drugs have divers parts; as that opium hath a stupefying part, and a heating part; the one moving sleep, the other a sweat following; and that rhubarb [root] hath purging parts, and astringent parts, &c. But this whole inquisition is weakly and negligently handled.[49]

'Purging medicines', Bacon writes elsewhere, 'for the most part, have their purgative virtue in a fine spirit; as appeareth by that they endure not boiling without much loss of virtue'. Once again, instead of reading these remarks 'metaphorically', we can tie them directly to his medicinal self-experiments. In the *Sylva sylvarum* he reports:

> Rhubarb hath manifestly in it parts of contrary operations: parts that purge, and parts that bind the body: and the first lie looser, and the latter lie deeper: so that if you infuse rhubarb for an hour and crush it well, it will purge better, and bind the body less after the purging, than if it stood twenty-four hours. This is tried [i.e. I have tried]. But I conceive likewise, that by repeating the infusion of rhubarb several times, ... letting each stay in but a small time, you may make it as strong a purging medicine as scammony.[50]

Once we begin to pay attention, we note that the language Bacon uses for the alterations to be effected by specific dosings with medicaments is identical to that in these ostensibly more 'philosophical', less 'medical' passages about transmission of thoughts and feelings. So it is not surprising to find that some of Bacon's clearest descriptions of the way the movement of animal spirits is the direct cause of mental states comes in his remarks on opium-taking, and material cause and effect in drug-therapy. For example, in the *History of Dense and Rare*:

> It may be with reason doubted whether opium and narcotics produce stupefaction by potential cold [the subject under consideration in this section], or by the flight of the [animal]

[49] *Sylva sylvarum*, Century I, 98; Spedding et al. 1857–1874, 2: 381.

[50] *Sylva sylvarum*, Century I, 20–21; Spedding et al. 1857–1874, 2: 345.

> spirits. For opium, from its strong smell, its bitterness, its sudorific power, and other signs, seems to have hot parts. But as it emits a vapour unfriendly and hateful to the spirits, it puts them to flight on all sides, whereby they are coagulated and suffocated [and the drug-taker goes into a stupour].[51]

And this applies, too, in the case of therapeutics which alter states of mind, where Bacon consistently talks in terms of the 'vapour' from the drug directly impacting on the 'spirits' in the brain:

> Opiates may not only be taken through the mouth, but likewise inhaled in the form of smoke; but it should be such as not to excite the expulsive faculty too strongly, nor draw out the humours, but only to work upon the spirits within the brain for a short time. Wherefore a suffumigation of tobacco, lign-aloes, dried leaves of rosemary, and a little myrrh, inhaled in the morning through the mouth and nostrils, would be very beneficial. ... Some medicines have a degree, weak and secret, and therefore safe, of opiate virtue. These impart a slow and abundant vapour, but not malignant, as opiates do. And hence they do not put the spirits to flight, but yet they collect and somewhat thicken them.[52]

So Bacon's animal spirits are vapours circulating inside the body, which can be altered – thereby altering states of mind – by applying *medicinal* vapours, ingested, smoked or inhaled. And here he agrees with another notable seventeenth-century physic-taker, Robert Boyle.

Boyle too disparages contemporary medicine for giving up on effects which cannot be witnessed from outside the body, and maintains that so-called 'occult' (hidden, as opposed to 'manifest') properties are perfectly reconcilable with a strictly mechanical philosophy:

---

[51] Spedding et al. 1857–1874, 5: 303. See also the section entitled 'The operation upon the spirits, that they may retain their youth and renew their vigour' (a section preoccupied with the efficacy of opiates and nitre) (ibid., 5: 268–283). Also: 'As to the motions corporal within the inclosures of bodies, whereby the effects ... pass between the spirits and the tangible parts (which are arefation, colliquation, concoction, maturation, &c.,) they are not at all handled. But they are put off by the names of virtues, and natures, and actions, and passions, and such other logical words.' (*Sylva sylvarum*, Century I, 98; Spedding et al. 1857–1874, 2: 382.)

[52] Spedding et al. 1857–1874, 5: 272. Again: 'Remembering what I have said before concerning opiates, nitre, and the like, which so powerfully condense the spirits, I do not think it would be amiss if once a fortnight three or four grains of castor be taken in morning broth, with a little angelica seed and calamus aromaticus. For these both strengthen / the brain, and excite I that density of the substance of the spirits (which is so necessary for longevity) vivacity and vigour of motion' (ibid., 5: 299–300). One of the clearest statement of such a mechanistic understanding of the mind/body relationship, brokered by the animal spirits, comes in the final section of the *History of Life and Death*, entitled 'Provisional rules concerning the duration of life and the form of death', which is taken up in its entirety with discussion of how to optimise the workings of the spirits within the human body. Under the rubric, 'Youthful spirits introduced into an old body may shortly turn back the course of nature' Bacon explains: 'The nature of the spirits is as it were the master-wheel [rota = cog/clockwork] which turns the other wheels in the body of man; and therefore in the intention of longevity it ought to stand first. Moreover there is an easier and more expeditious way of altering the spirits than the other parts. For the operation upon the spirits is twofold; the one by aliment, which is slow and as it were circuitous; the other (itself likewise twofold) which is sudden, and goes at once to the spirits, – namely, by vapours or by the affections.' (ibid., 5: 330.)

> That scarce sensible quantities of matter, having once obtained access to the mass of the blood [by injection] … may give such a new and unnatural impediment or determination to the motion of the blood, as to discompose … its texture … (as a spark of fire reduceth a whole barrel of gunpowder …) need be manifested by nothing but the operations of such poisons, as work not by any of those (which physicians are pleased to call) 'manifest qualitites'. … The known effects of a very small quantity of opium, or of arsenick, of the scarce discernable hurt made by a viper's tooth, and especially of the biting of a mad dog …, are sufficient to evince what we [have] proposed.[53]

Boyle shared Francis Bacon's liking for dosing himself with wormwood. In August 1649, 6 months after the execution of Charles I, when a good number of elite gentlemen in England were suffering from nervous illness brought on by political events, Boyle (who though his family home was in Ireland, was currently living on his estate at Stalbridge in England) was recuperating from an ague and 'drawing a quintessence of wormwood (for my particular use)'.[54] Back in Ireland in 1653, his physician William Petty (also physician to Cromwell), wrote Boyle a stern letter about self-diagnosis and prescription. Boyle is reproved for 'your apprehension of many diseases, and a continual fear, that you are always inclining or falling into one or the other'. This fear, Petty goes on, is itself a disease – hypochondria –, 'incident to all, that begin the study of diseases'. And he warns Boyle of the dangers of

> "practising upon yourself with medicaments (though specificks) not sufficiently tried by those, that administer or advise them."[55]

In a longer version of this paper I would want to argue that Boyle's interest in experiments in the air pump, combined with his dissections and vivisections concerning the flow of blood through the heart and the lungs, are fundamentally concerned with the relation between body and spirit.[56]

As I draw towards my conclusion, I turn, in the light of my discussion here, to the finale to Bacon's own life – the events leading up to his death. In spite of the lingering appeal of Aubrey's apocryphal 'chicken-stuffing' story, Alan Stewart and I showed in our biography, *Hostage to Fortune*, that there is evidence to suggest that Bacon died as a result of physic-taking, in the form of inhaled fumes of heated nitre (saltpetre). You will, of course, all be familiar with the story that Bacon died as the result of a chill he caught, having alighted from his carriage to stuff the cavity of a chicken with snow, on Highgate Hill (in April). That story (we show) is a 'Chinese whispers' version of an account Bacon himself gave of the fatal incident. Forced by

---

[53] Hunter and Davies 1999, 2: 170–83; 5, 77; cit. (in the contrary sense!) Hutchison 1997, 100–101.

[54] Cit. Maddison 1963, 113.

[55] Cit. ibid., 116.

[56] For suggestive articles about Boyle and medical experiment see Hunter 1999; Dewhurst 1962; Kaplan 1993, 94–5; Schaffer 1998, 94–100. See also the entry in the Oxford physician John Ward's diary, as reported by Frank: 'in early 1659 he recorded a note about Boyle's experiment of syringing chemical emetics into dogs' veins', also 'In the early spring in 1659 he noted that the technique of injecting opium and crocus metallorum into a dog's veins was first practiced by Christopher Wren and Robert Boyle at Oxford and London' (Frank 1974, 155, 156)

violent vomiting to take refuge in the Highgate house of the earl of Arundel, he wrote a letter (before he realised how seriously sick he was), to explain why (in the earl's absence) he had had without warning taken advantage of the hospitality of his household. The letter was printed in 1660 by Tobie Matthew in his *Collection* of letters:

> My very good Lord, – I was likely to have had the fortune of Caius Plinius [Pliny] the elder, who lost his life by trying an experiment about the burning of the mountain Vesuvius. For I also was desirous to try an experiment or two, touching the conservation and induration [strengthening] of bodies. As for the experiment itself, it succeeded excellently well; but in the journey (between London and Highgate) I was taken with such a fit of casting [vomiting], as I knew not whether it were the stone, or some surfeit, or cold, or indeed a touch of them all three. But when I came to your Lordship's house, I was not able to go back, and therefore was forced to take up my lodging here, where your housekeeper is very careful and diligent about me; which I assure myself your Lordship will not only pardon towards him, but think the better of him for it. For indeed your Lordship's house was happy to me; and I kiss your noble hands for the welcome which I am sure you give me to it, etc. I know how unfit it is for me to write to your Lordship with any other hand than mine own; but in troth my fingers are so disjointed with this fit of sickness, that I cannot steadily hold a pen.[57]

Read with an ear for the element of self-ironising which Bacon intended (it was, of course, written before Bacon had any idea that his illness would prove fatal), this letter is remarkably clear and specific. Bacon had been conducting self-experiments for the prolongation of life in living bodies involving inhalation of vapours (like Pliny on the slopes of Vesuvius) in central London. On the return journey to St Albans he was overcome by violent vomiting and forced to break his journey at the nearby home of the earl. He died there a couple of days later.

We also proposed that the substance Bacon was inhaling was nitre (saltpeter) on the strength of a passage from the *Sylva sylvarum* which runs as follows:

> There be two things which (inwardly used) do cool and condense the spirits; and I wish the same to be tried outwardly in vapours [inhaled]. The one is nitre, which I would have dissolved in Malmsey, or Greek wine, and so the smell of the wine taken; or if you would have it more forcible, pour of it upon a fire-pan, well heated, as they do rose-water and vinegar. The other is the distilled water of wild poppy [i.e. opiates], which I wish to be mingled, at half, with rose-water, and so taken with some mixture of a few cloves in a perfuming-pan.[58]

At the time of writing our biography, we were unable to explain exactly what the effects of inhaling warmed potassium sulphate would have been expected to be. Recently, however, we discovered that here, as so often, Bacon was ahead of his time scientifically (in this case, his informant is likely to have been the Dutch scientist, resident in England from 1604 – with a brief interlude in Prague –, Cornelis Drebbel). When lightly heated, saltpetre gives off pure oxygen, which makes the self-experimenter/physic-taker feel euphoric and (to quote Joseph Priestly) 'wise'.

[57] FB to Arundel, n.d. Matthew, *Collection*, 57–8; *LL* 7: 550. For discussion see Jardine and Steward 1998, 504–5.

[58] *Sylva sylvarum*, Century 10, Spedding et al. 1857–1874, 2: 650.

Priestly is credited with discovering this in 1784, (in his case by heating Mercuric Oxide). He predicted that this new 'pure air' would become a luxury drug.

Since in Bacon's day saltpetre was obtained by collecting it in its naturally occurring form as bird-droppings and animal dung, fermented under particular conditions in situ and then 'mined' by saltpeter-men, it was frequently impure, though pure enough for its primary purpose as a key ingredient in gunpowder. It was, one imagines, one of these impurities which produced the caustic fumes which first overcame Bacon ('I was likely to have had the fortune of Caius Plinius the elder, who lost his life by trying an experiment about the burning of the mountain Vesuvius'), and then corroded his lungs, causing death by suffocation.

I do not, however, intend to close here with the misfortune of Bacon's 'bad batch' of nitre. Rather, I take it as a parable for our need to reconsider the relationship in Bacon's work between his philosophical and experimental medical activities.

The daily dosing carried out by scientific *virtuosi* like Bacon, Boyle and Hooke constituted some of the most systematic 'experimenting' in their scientific repertoire. It was certainly some of the most 'hands-on' experimenting Bacon and Boyle did (since they customarily used laboratory assistants for other types of experimental work). As self-experiment, it allowed the experimentalist unique and direct access to the effects of the substances administered, including, importantly, effects on mood and mental alertness. My suggestion is that these were experiments, and lines of inquiry, which engrossed seventeenth-century scientists, and along the way shaped subsequent discursive productions of the thinking subject.

My further suggestion is that in the surviving records of such experimenting on one's own body we can observe individuals like Bacon, Boyle and Hooke, deliberately setting out to repeat pleasurable, or intellectually productive moods and states of mind, by applying the same 'physic' on successive occasions (often, daily). We are watching, therefore, theories of the working of the human mind being developed by key members of an urban culture (London; interregnum Oxford), who based those theories on their own regular, long-term use of self-administered, mind-altering substances. This really is not to overstate the case – leaving aside opiates and narcotics, purgatives like *sal ammoniac* and *crocus metallicus* have known stimulant properties in the kind of doses they were taking. (By the way, the whole thing is remarkably systematically conducted: the scientists' daily routines are carefully designed around timed doses, as well as access to suitable facilities at appropriate times for the vomiting, urinating and excreting which followed some hours after the physic had been administered!)

Finally, let me return to Robert Hooke – avid Baconian, and emulator of Baconian scientific practice –, who provided me with my opening epigram. In the preface to *Micrographia*, Hooke likened the process whereby hand and eye, memory and mind, arrived at a clear understanding of nature to the circulation of the blood:

> So many are the links, upon which the true Philosophy depends, of which, if any one be loose, or weak, the whole chain is in danger of being dissolv'd; it is to begin with the Hands and Eyes, and to proceed on through the Memory, to be continued by the Reason; nor is it to stop there, but to come about to the Hands and Eyes again, and so, by a continual passage round from one Faculty to another, it is to be maintained in life and strength, as much as the

> body of man is by the circulation of the blood through the several parts of the body, the Arms, the Feet, the Lungs, the Heart, and the Head.[59]

Here, I think, Hooke takes up Bacon's argument about animal spirits, in the context of his own generation's (and above all his own) heightened awareness of the mental consequences of habitual 'physic'. The iterative process of sharpening the connections between physical and mental faculties (Hand and Eye, Memory and Reason/Mind) is like 'the circulation of the blood through the several parts of the body', not to mention the circulation by way of that same blood-stream of substances deliberately entered into it by mouth, inhalation or injection.

Self-dosing with substances designed to produce euphoria and wisdom was, for Bacon and his disciples, routine – an intellectual therapy. It formed part of a considered programme for modifying the animal spirits, to extend and improve the faculties of body and mind. In other words, for the ailing early modern natural philosopher, ongoing therapy for chronic symptoms went hand in hand with enhancing the natural faculties, as one step along the path to knowledge.

Bibliography

Beier, Lucinda McCray. 1989. Experience and experiment: Robert Hooke, illness and medicine. In *Robert Hooke: New studies*, ed. M. Hunter and S. Schaffer, 235–252. Woodbridge/Suffolk: Boydell Press.

Clark, George. 1964–1966. *A history of the Royal College of Physicians of London*, 2 vols. Oxford: Clarendon Press.

Crowther, James. 1982. *Founders of British science*. Westport: Greenwood Press.

Debus, Allen G. 1965. *The English paracelsians*. London: Oldbourne.

Dewhurst, Kenneth. 1962. Locke's contribution to Boyle's researches on the air and on human blood. *Notes and Records of the Royal Society* 17: 198–206.

Findlen, Paula. 1997. Francis Bacon and the reform of natural history in the seventeenth century. In *History and the disciplines: The reclassification of knowledge in early modern Europe*, ed. D.R. Kelley, 239–261. Rochester: University of Rochester Press.

Frank, Robert G., Jr. 1974. The John Ward diaries: Mirror of seventeenth century science and medicine. *Journal of the History of Medicine and Allied Sciences* 29: 147–179.

Gouk, Penny. 1999. *Music, science and natural magic in seventeenth-century England.* New Haven: Yale University Press.

Gunther, Robert T. 1926. *Early medical and biological science*. Oxford: Oxford University Press.

Halley, Edmond. 1937. In *Correspondence and papers of Edmond Halley*, ed. E.F. MacPike. London: Taylor and Francis.

Hooke, Robert. 1665. *Micrographia*. London: Printed by Jo. Martyn and Ja. Allestry.

———. 1726. In *Philosophical experiments and observations of the late eminent Dr. Robert Hooke*, ed. W. Derham. London: Printed by W. and J Innys.

———. 1935. In *The diary of Robert Hooke MA MD FRS 1672–1680*, ed. H. Robinson and W. Adams. London: Taylor and Francis.

Hunter, Michael. 1999. Robert Boyle (1627–1691): A suitable case for treatment? *British Journal for the History of Science* 32: 261–275.

[59] Hooke 1665, preface.

Hunter, Michael and Edward B. Davis. 1999–2000. *The works of Robert Boyle*, 14 vols. London: Pickering & Chatto.

Hunter, Lynette, and Sarah Hutton, eds. 1997. *Women, science and medicine 1500–1700*. Gloucestershire: Sutton Publishing.

Hutchison, Keith. 1997. What happened to occult qualities in the scientific revolution? In *The scientific enterprise in early modern Europe: Readings from isis*, ed. Peter Dear, 86–106. Chicago: Chicago University Press.

Jardine, Lisa. 2002. *On a grander scale: The outstanding life of Sir Christopher Wren*. London: Harper Collins.

Jardine, Lisa, and Michael Silverthorne, eds. 2000. *Francis Bacon: A new organon and other writings*. Cambridge: Cambridge University Press.

Jardine, Lisa, and Alan Stewart. 1998. *Hostage to fortune: The troubled life of Francis Bacon*. Gollanc: Phoenix.

———. 2000. Editing a hero of modern science. In *Books and the sciences in history*, ed. Marina Frasca-Spada and Nicholas Jardine, 354–368. Cambridge: Cambridge University Press.

Kaplan, Barbara. 1993. *'Divulging of useful truths in Physick': The medical agenda of Robert Boyle*. Baltimore: The Johns Hopkins University Press.

Maddison, R.E.W. 1963. Studies in the life of Robert Boyle FRS: Part VI. The Stalbridge period, 1645–1655, and the invisible college. *Notes and Records of the Royal Society* 18: 104–124.

Multhauf, Robert. 1954. Medical chemistry and "the Paracelsians". *Bulletin of the History of Medicine* 28: 101–126.

Parfitt, George, ed. 1975. *Ben Jonson: The complete poems*. Harmondsworth: Penguin Education.

Plant, Sadie. 1999. *Writing on drugs*. London: Faber.

Porter, Roy, and George S. Rousseau. 1998. *Gout: The patrician malady*. New Haven: Yale University Press.

Schaffer, Simon. 1992. Self evidence. *Critical Inquiry* 18: 327–362.

———. 1998. Regeneration: The body of natural philosophers in restoration England. In *Science incarnate*, ed. Lawrence and Stephen Shapin, 83–120. Chicago: University of Chicago Press.

Siraisi, Nancy G. 1987. *Avicenna in renaissance Italy: The canon and medical teaching in Italian universities after 1500*. Princeton: Princeton University Press.

Spedding, James, Robert L. Ellis and Douglas D. Heath, eds. 1857–1874. *The works of Francis Bacon*, 14 vols. London: Longmans.

Chapter 4
Francis Bacon's Body and His Experiments on the Prolongation of Life

James Everest

Abstract This chapter pursues the thrust of the volume with specific regard to the life and work of the natural philosopher Francis Bacon (1561–1626). Firstly, it puts Bacon's episodes of ill-health front and centre in an account of his political career; secondly, it places his philosophical work on life and death in this obvious, but under-regarded context; and thirdly, it sets his advocacy of experimentation against the backdrop of his life-long practice of self-medication, again an obvious but under-regarded context. The joining of the dots going on in this piece results in a small but important step in the ongoing efforts of historians of science to reconnect the history of medical empiricism with the history of experimental science.

Keywords Francis Bacon · Health · Prolongation of life · Experimentation · Medical empiricism

In 1608, Francis Bacon (1561–1626) began the long summer break from his responsibilities in the law courts by taking stock of his life. He had recently obtained, after 19 years of waiting, the post of Clerk of the Star Chamber, a development that may have had something to do with the urge to gather in one place ideas and observations relating to his various personal and professional projects. His notebook, entitled the *Commentarius solutus*, is now lodged in the British Library.[1] One

My thanks to Angus Gowland, Alan Stewart and Gideon Manning for their thoughts on this essay and to Mordechai Feingold and Jed Buchwald for hosting me at the conference at Caltech where the ideas contained within it were first articulated. The inspiration for the piece, as for so much, was Lisa Jardine.

[1] For the context, see Jardine and Stewart 1998, 299. For a discussion of the manuscript – British Library Additional MS 27278 – see Vine 2011.

J. Everest (✉)
University of Birmingham, Birmingham, UK

G. Manning (ed.), *Testimonies: States of Mind and States of the Body in the Early Modern Period*, Archimedes 57,
https://doi.org/10.1007/978-3-030-39375-5_4

particularly striking comment, towards the front of the manuscript, begins: "After a Maceration taken in y^{e} morning and working littell I tooke a glister about 5 of clocke to draw it down better".[2] We know from his chaplain William Rawley's *Life*, published in 1657, that Bacon was in the habit of taking medicinally a "maceration of rhubarb, infused into a draught of white wine and beer mingled together for the space of half an hour [i.e. rhubarb soaked in this concoction for this time]", which he would consume "once in six or seven days".[3] Suggestions for ingredients to be added to the mix, including the flower elecampane and the byproduct of winemaking known as 'cream of tartar', appear in a collection of notes printed in 1679 as his *Medical Remains*.[4] The same collection features a recipe for a "cooling clyster [an enema] to be used once a month, after the working of the maceration is settled".[5]

On this occasion, however, something went wrong. Bacon notes that his enema "wrought but slowly, neyther did I find that lightness and cooling in my sydes which many tymes I doe, but soon after I found a symptome of melancholy".[6] He took remedial action, in the form of three purgative pills, which effectively "wrought within 2 howres without gryping [bowel pain] or vomite and brought much of ye humor [presumably black bile] sulphureous and feetide". The narrative then draws to a happy conclusion: "though my medicine was not fully settled I made a light supper without wyne, and fownd myself light and at peace after it".[7]

What are we to make of this passage? One obvious approach is to view it as an insight into the life of the man who lies behind some of the most famous pieces of scientific writing ever produced: this tale of concoctions and purging provides a new way of looking at an intellectual figure who, in the relative privacy of his personal notebook, turns out to have been fascinated by the sulphurous and fetid humours of his own body.[8] In the introduction to her piece for this volume, Lisa Jardine goes a step further, using the passage as an encouragement "to reconnect health and thought", two domains that, as she comments, "were not understood as separate in the seventeenth century".[9] From this perspective, the significant feature of the account is the fact that 'melancholy' appears as both a physical and a mental malaise, with Bacon considering both aspects as part of the same problem.[10] Furthermore, his intervention can, as Jardine suggests, be viewed as a manifestation of his experimental approach to natural philosophy: he describes the state of affairs, carefully documents his actions and concludes by noting the outcome. In the margin, he even

[2] *Commentarius solutus*, in Bacon 1861–74, vol. 4, 53.

[3] Rawley 1857–59, vol. 1, 17.

[4] For an introduction to the notes of which the *Medical Remains* are a part, see Rees 1981, 380.

[5] Bacon, *Medical Remains*, 833.

[6] Bacon, *Commentarius solutus*, 53–54.

[7] *Ibid.*, 54.

[8] Lisa Jardine and Alan Stewart quote this passage at length in their biography of Bacon; see 1998, 300–1.

[9] See chap. 3.

[10] For the place of melancholy in early seventeenth-century intellectual culture, see Gowland 2011.

speculates as to what might have caused the unexpected turn of events: "Note there had been extreme heates for x dayes before and I had taken littell or no phisicke".[11]

There is, however, some further 'reconnecting' going on in Jardine's piece. On the surface, the goal is clearly to bring together two parts of early modern thought, to show that, for philosophers of this period, how they thought about how they felt was on a spectrum with how they thought about how they thought. Under the surface, however, the essay does something else. In highlighting the importance that her subjects ascribed to their "state-of-mind-altering experiments", Jardine seems to be reconnecting feeling and thought in a more profound way.[12] To my mind at least, her piece manages to throw new light not simply on how seventeenth-century philosophers thought about how they felt, but also on how they felt about how they thought. In a passage that Jardine quotes from the diary of Robert Hooke, for example, the Royal Society's Curator of Experiments clearly appreciates the clear-headedness with which he has been able to spend a night reflecting on water-clocks, quadrants, scotoscopes, *etc.*, a state that resulted from dosing himself up with a mixture of toxic substances.[13] Another significant aspect of the passage in the *Commentarius solutus*, meanwhile, is the fact that Bacon judges his actions based on how he *feels* as a result: he diagnoses the failure of his enema with reference to the absence of expected sensations of lightness and cooling in his sides; his purgative pills succeed when, after supper, he feels light and at peace.

Here, my intention is to pick up on this connection between thought and feeling, a connection that I am inspired to pursue both by Lisa Jardine's contribution to this volume and by the attention paid to the emotions in recent special issues of the history of science journals *Isis* and *Osiris.*[14] The Francis Bacon that I will be describing led a life that was punctuated by episodes of ill health. He was, moreover, particularly conscious of his suffering: this Bacon was a man of ferocious ambition, who knew how it felt to be let down by his body. His experience of being embodied, I will suggest, is an obvious but hitherto under-regarded backdrop for an important strand of his natural philosophy, his writing on life and death. The significance of his body, meanwhile, may not be limited to his choice of natural philosophical subject-matter; I will go on to suggest that it may also have played a role in his approach to natural philosophical questions. Famously, Francis Bacon advocated the advancement of learning by means of information gained from experiments. Less famously, he also pursued his own sophisticated course of experimental investigations. My temptation here is to join the dots and suggest that his belief in the value of experimental knowledge may have owed something to his life-long practice of medical self-experimentation.

---

[11] *Commentarius solutus*, in Bacon 1861–74, vol. 4, 53.

[12] The quote is from chap. 3.

[13] *Ibid.*

[14] *Isis* 100, no. 4 (2009) and *Osiris* 31, no. 1 (2016).

4.1 Bacon's Body

It seems fair to say that Bacon's body has not generally been a subject of much interest for Bacon scholars, although it was clearly of considerable interest for him. The exception is Lisa Jardine and Alan Stewart's biography, which makes clear that, for much of his life, this prominent statesman and famous natural philosopher was also a frail, puny, languishing creature.[15] Both he and his brother Anthony were, as they highlight, sickly children, who as young teens at Trinity College, Cambridge, in the 1570s exhausted the funds of their tutor John Whitgift with expensive purchases, including unusually large quantities of coal in summer.[16] And physical frailty appears to have remained with them into adulthood: on 24 May 1592, their mother – the formidable Lady Anne Bacon – wrote to Anthony:

> I verely think your brother's weake stomack to digest hath ben much ~~begun~~ caused and confirmed by untimely ^late^ going to bed and then musing *nescio quid* [I know not what] when he shuld slepe and then in conseqwent by late rising and long lyeing in bed. Wherby his men are made slewthfull and himselff continuall syckly[17]

To set this maternal lament in some context, in May 1592 Francis was thirty-one. He had studied law at Gray's Inn and had what looked like a promising political career in front of him, under the patronage of the Earl of Essex.[18] In the same year, he sent a letter to his uncle, the Queen's chief minister Lord Burghley, which provides a sense of his ambition. The letter is famous for the first statement of what would become a revolutionary program for the advancement of learning – Bacon writes of having "as vast contemplative ends, as I have moderate civil ends"[19] – but what is less often mentioned is the fact that the entire first half of the letter is taken up with an account of those supposedly 'moderate' civil ends. Indeed, the very purpose of sending it is to stress his desire for a position in the Queen's service: "I ever bare a mind", he writes, as if previous expressions of the same plea had fallen on deaf ears, "to serve her Majesty". The re-iteration of his willingness to serve follows immediately a passage that appears designed to re-assure his correspondent: "My health, I thank God, I find confirmed; and I do not fear that action shall impair it, because I account my ordinary course of study and meditation to be more painful than most parts of action are".[20] The claim looks fanciful – those who engage in a lifetime of study and meditation are not generally renowned for a resulting robust-

[15] For their references to Bacon's health, see above, n. 8, and Jardine and Stewart 1998, 37, 136, 172, 451, 489, 495 and 520.

[16] See *Ibid.*, 36–37.

[17] Allen 2014, 105. The carets are the editor's means of representing an intralinear insertion in the manuscript. For an overview of Lady Bacon's life and a sense of her character, see the introduction, 1–45.

[18] See Jardine and Stewart 1998, 115–17 and 121–25.

[19] Bacon 1861–74, vol. 1, 109.

[20] *Ibid.*, 108.

ness of constitution – but the purpose of advancing it is clear: Bacon is seeking to prevent a reputation for ill health from impinging on his career prospects.

In the following year, 1593, the prestigious position of Attorney-General fell vacant and Bacon, perhaps vaingloriously, launched a campaign.[21] His forwardness seems to have been greeted with apprehension. In an undated letter to the Queen, he seeks to allay the suspicion that his action is motivated by profit: "I affected myself to no great matter, but only a place of my profession, such as I do see divers younger in proceeding to myself, and men of no great note, do without blame aspire unto".[22] He continues: "But I will not wrong mine own good mind so much as to stand upon it now, when your Majesty may conceive I do it but to make my profit of it".[23] Bacon was habitually short of money, so questions about his motivation could easily have arisen.[24] Concerns about his suitability may also, however, have had roots elsewhere. In August, Lady Bacon wrote a letter, now lost, to her brother-in-law Burghley on his behalf. For the most part, Burghley's response is carefully-phrased flummery: "thoughe I am of lesse power to doe my friends good then the worlde thinketh", he replies, "yet they shall not want the intention to doe them good".[25] The response, however, begins with a comment that is almost certainly telling: "for your sons, I think your care of them is no less than they deserve, being so qualified in learning and virtue as if they had a supply of more health they wanted nothing".[26]

If anxiety about Bacon's health was a potential stumbling block in his campaign for political advancement, he did not help his cause when, on a royal commission in July 1594, he had to interrupt his journey northwards, due to illness. Forced to inform the Queen, he writes: "As I do acknowledge a providence of God towards me that findeth it expedient for me *tolerare jugum in juventute meâ* [to bear a yoke in my youth], so this present arrest of me by his divine Majesty from your Majesty's service is not the least affliction that I have proved".[27] This letter comes at a particularly miserable juncture in the career of a man who, at this point, was a long way from the powerful statesman and well-known natural philosopher that he was to become: although he does his best, there is not really much positive spin that can be put on the admission that he is liable to fall ill while on his sovereign's service.

By this point, Bacon had already lost out on the Attorneyship to the older and more experienced lawyer Edward Coke, appointed earlier that year. By the following summer, his chances for the more humble position of Solicitor-General, which had opened up when Coke was promoted, were also looking unlikely, with the position eventually going to another older and more experienced lawyer, Thomas

---

[21] See Jardine and Stewart 1998, 146–53.

[22] Bacon 1861–74, vol. 1, 240–41.

[23] *Ibid.*, 241.

[24] Bacon's spendthriftiness is another feature of Jardine and Stewart's biography; see Jardine and Stewart 2008, 48, 191, 198–208, 258, 275, 443, 463–64, 487 and 511–12.

[25] Allen 2014, 147.

[26] *Ibid.*, 146.

[27] Bacon 1861–74, vol. 1, 305.

Fleming.[28] The twofold frustration of his hopes, perhaps unsurprisingly, seems to have taken its toll on his health. In a letter sent on 5 August, Lady Bacon writes to Anthony: "I am sory your brother with inwarde secret greeff hindreth his health. Every body sayth he loketh thine and pale". She continues: "I had rather yee both with God his blessed favour had veri goode health and well owt of debt, then any office yet".[29]

Bacon's failure to obtain political advancement in this period has traditionally been ascribed to the Queen's anger at a speech delivered in Parliament in early 1593, in which he criticised the Crown's request for additional funds (in a debate over what is known as the 'triple subsidy').[30] In a study of Bacon's political career, Julian Martin has suggested that his "youthfulness and relative lack of experience" are more probable causes for the Attorneyship going elsewhere and has attributed the decision on the Solicitorship to Elizabeth's "determination to maintain her autonomy" in the face of "Essex's ceaseless and public demands for his friend's appointment".[31] The situation is likely to have been multifaceted, with the evidence presented here suggesting that Bacon's health problems may also have had a part to play. If so, going over this episode is valuable not simply for the reminder that poor health evidently dated to an early point in his life. The episode is also valuable for the sense that his health may have acted as a limiting factor for his career: a man with an obviously vaulting sense of ambition, he may have had experience of the extent to which the fulfilment of his desires was subject to the limits of his body.

4.2 The Prolongation of Life

Bacon's body did not hold him back forever; under James I, himself a man with a far from robust disposition,[32] he gradually gained political responsibilities, including the role of Solicitor-General in 1607.[33] At about the same time, he also began developing his thoughts on natural philosophical questions. His statement about the value and current state of knowledge, *The Advancement of Learning*, was published

[28] See Jardine and Stewart 1998, 168–72.

[29] Allen 2014, 229.

[30] See, for example, Jardine and Stewart 1998, 141–44.

[31] Martin 1992, 98 and 100.

[32] A classic description is Sir Anthony Weldon's: "His Tongue [was] too large for his mouth, which ever made him speak full in the mouth, and made him drink very uncomely, as if eating his drink, which came out into the cup of each side of his mouth [...] His Legs were very weake, having had (as was thought) some foul play in his youth, or rather before he was born, that he was not able to stand at 7 years of age, that weaknesse made him ever leaning on other mens shoulders" (Weldon 1651, M2^r). Weldon may have had reason to slander James, having possibly been dismissed from court (see Marshall and Kelsey, 'Weldon, Sir Anthony' accessible through www.oxforddnb.com), but his description of the King's gait tallies with other accounts, including that of Theodore de Mayerne, the French physician who treated James late in life; see Stewart 2003, 38.

[33] See Jardine and Stewart 1998, 296.

2 years into the new King's reign, in 1605,[34] and a number of draft pieces dating to this period survive.[35] At some point in the 1610s he began a manuscript work with a dramatic title – *De vijs mortis* (the ways of death) – that is his earliest known foray into a field of study that would remain important for him: the prolongation of life.

The piece in question is more positive than its gloomy title perhaps suggests; in fact it opens with a startling claim: "Anything that can be constantly fed, and by feeding be wholly restored is like the vestal flame, potentially everlasting".[36] Bacon provides a brief overview of current thinking in the field – and why it is wrong – before stating what he thinks is the fundamental novelty of his approach: "to avoid making a hash of the investigation, anyone who tries to examine the deterioration of an animal, and investigate its causes and cures (if any there be) or the means at least of delaying it, ought to start by looking at the animal first as an inanimate thing which does not take on food, and only then separately as an animate thing which does".[37]

Over the same period, his career continued to gather steam. In 1613, he was finally made Attorney-General; in 1617, Lord Keeper of the Privy Seal; in 1618, Lord Chancellor.[38] In 1621, however, he fell dramatically from grace; after confessing to have taken bribes, he was barred from Parliament and banished from court.[39] He spent much of the remaining 5 years of his life at his estate in Gorhambury, near St Albans, where he turned his attention to a number of intellectual projects, including the publication of a series of natural histories.[40] He had announced a turn to natural history at the end of the volume, published in 1620, that introduced the world to the six-part plan for his project to reform learning, the 'Great Instauration'. To some extent, he had already met the goal for part one, an overview of the current state of knowledge, with the publication in 1605 of the *Advancement of Learning*. (This period also saw the translation into Latin and dramatic expansion of the work, published in 1623 as the *De augmentis scientiarum*.[41]) The main body of the 1620

[34] *The Advancement of Learning*, in Bacon 2000a, 1–193.

[35] The pieces will be included in volume five of *The Oxford Francis Bacon*.

[36] *Quod ali perpetuó potest, et alendo restitui in integrum, id potentiâ æternum est, tanquam ignis vestalis* (*De vijs mortis*, in Bacon 1996, 270, translated on 271). Graham Rees's edition attempts to replicate typographically the palimpsestic nature of the manuscript. For the sake of clarity, I have not followed suit here.

[37] *ut inquisitio non confundatur, qui exolutionem Animalis intueri, eiusque causas atque remedia si quæ sunt, aut saltem moras, perscrutari conatur, tale initium sumere debet ut contemplatur, primum, Animal tanquam inanimatum, atque inalimentatum; posteá veró ac seorsim tanquam animatum et alimentatum* (*De vijs mortis*, in Bacon 1996, 274, translated on 275).

[38] See Jardine and Stewart 1998, 340–41, 393 and 420.

[39] For a description of the case, see Jardine and Stewart 1998, 444–69. For Bacon's confession, see Bacon 1861–74, vol. 7, 252–62.

[40] Bacon's natural histories have been the object of some attention in recent years; see, for example, the collection of articles published in a 2012 special issue of *Early Science and Medicine*, as well as Manzo 2009; Jalobeanu 2010 and 2015, 199–250.

[41] *De augmentis scientiarum*, in Bacon 1857–59, vol. 1, 415–840, trans. vols. 4 and 5, 273–498 and 3–123.

volume is taken up with part two, the method, named after Aristotle's 'Organon',[42] but at the end of the volume he includes a brief introduction to part three, natural history, with a powerful statement of its importance:

> if all the brilliant people of all ages past and to come were to join together, if the entire human race had devoted or were to devote itself to the work of philosophy, and if one had a whole world past and future overrun with academies, colleges, and the schools of learned men, they could still not have made and cannot make any headway in philosophy and the sciences worthy of the human race without a natural and experimental history of the kind for which I shall now give instructions[43]

Bacon's first contribution to natural history – a volume entitled the *Historia naturalis et experimentalis* – was published in 1622. After a dedicatory epistle to Prince Charles, he lists six titles that he (somewhat optimistically) intends to publish over the following 6 months.[44] The first, the *Historia ventorum* (a natural history of the winds), was included in the same volume,[45] but the second, the *Historia densi & rari* (a natural history of density and rarity), never made it beyond a manuscript draft.[46] Instead, in 1623, Bacon chose to publish the work that he had originally placed last on his list, the *Historia vitæ & mortis*, explaining in his address to the reader: "Although I placed the History of Life and Death last on the list of the six monthly histories that I planned, I have decided to bring it forward and publish it in second place on account of the exceptional utility of the matter, a matter in which the slightest loss of time should be counted precious".[47]

Bacon's publication of a history of life and death as part of his Great Instauration serves, firstly, to emphasise the fact that there is nothing incidental about his work in the field: this is one of the two natural histories that he published and, as we have seen, he thought of natural history as essential for his project for the advancement of human knowledge. There is, however, perhaps another context in which we should view the publication of the work. In 1623, Bacon was 62, and his health was not improving. The previous winter had been particularly cold; in an undated letter, sent at some point in the spring to George Villiers, the Duke of Buckingham, he complains: "This extreme winter hath turned with me weakness of body into a state

---

[42] *Novum organum*, in Bacon 2004, 50–447.

[43] *Non si omnia omnium Ætatum ingenia coiuissent, aut posthâc coierint; Non si vniuersumgenus humanum Philosophiæ dedisset operam, aut dederit; Et Totus Terrarum Orbis nihil aliud fuisset, aut fuerit, quàm Academiæ, & Collegia, & Scholæ virorum doctorum; tamen absque tali, qualem nunc præcipiemus, Historiâ Naturali & Experimentali, vllos qui genere humano digni sint progressus, in Philosophiâ & Scientijs fieri potuisse, aut posse*, Bacon, *Parasceve ad historiam naturalem et experimentalem* (Bacon 2004, 450–52, translated on 451–53).

[44] *Historia naturalis et experimentalis*, in Bacon 2007, 6.

[45] *Historia ventorum,* in Bacon 2007, 18–131.

[46] *Historia densi & rari*, in Bacon 2000b, 1–34 and 35–169.

[47] *Cvm* Historiam Vitæ & Mortis, *inter* Sex Designationes Menstruas, *vltimo loco posuerimus; omninò hoc præuertere visum est, & secundam edere, propter eximiam rei vtilitatem; in quâ, vel minima temporis iactura, pro pretiosâ haberi debet* (Bacon 2007, 142, translated on 143).

that I cannot call health but rather sickness".[48] When, therefore, he describes the subject of his natural history as of such "exceptional utility" that "the slightest loss of time should be counted precious", it is tempting to ask: precious for whom?[49]

Much of the material in the *Historia vitae & mortis* is plundered from the *De vijs mortis*. Like the earlier work, it opens with the optimistic notion that the life of a living body can potentially be prolonged indefinitely.[50] It also features Bacon's statement of his novel approach: to think about animate bodies initially as if they were inanimate.[51] His claim to novelty is, as Graham Rees notes in his introduction to the *De vijs mortis* for the *Oxford Francis Bacon* edition of the work, not totally unjustified.[52] Meanwhile, although the idea of approaching animate bodies in this way may sound a little like the mechanistic thinking that would have such a powerful impact in subsequent decades, Bacon's conception of matter in fact involves the attribution of 'spirits' to all kinds of bodies, both animate and inanimate.

Large parts of the *De vijs mortis* and the *Historia vitæ & mortis* are taken up with theoretical discussions about the nature of these spirits – inanimate in the case of inanimate bodies, inanimate and vital in the case of animate ones – and how they interact with matter: in particular, the tendency of the inanimate spirits to destroy their animate hosts and the power of the vital spirits to preserve them. Such questions have accordingly occupied a central position in the scholarly writing on the subject.[53] It is, however, also worth highlighting some of Bacon's more practical corollaries. As part of his discussion of the destructive potential of the inanimate spirits in the *De vijs mortis*, for example, he states the aphorism: "Extrinsic spirits clinging loosely to a body corrupt the innate and immersed spirits of the thing, stimulate and make them ready to escape, and for that reason speed up the dissolution of the thing".[54] The "first rule of conservation" [*Canon conseruationis primus*] that follows is: "bodies ought to be exposed to the air as soon as they have been picked, and be put in a cool shady place where the sun's heat cannot disturb the inner spirits, but where the air can attract and remove the lighter spirits".[55]

---

48 Bacon 1861–74, vol. 7, 398.

49 As Richard Serjeantson has noted, the prolongation of life is a prominent theme in other works dating to the last years of Bacon's life, including the *De augmentis scientiarum* and the subject of his essay, the utopian fable *New Atlantis*; see Serjeantson 2002, 89–95.

50 *Historia vitæ & mortis*, in Bacon 2007, 144.

51 *Ibid.*, 148.

52 Rees 1996, lxix.

53 There is now a considerable scholarly literature on Bacon's understanding of spirits; for an overview, see Rees 1996, liv-lxix, which condenses material previously published as Rees 1977, 1984a, 1984b, 33–62. For the role of the spirits in the prolongation of life, see the works referenced below, nos. 58, 59 and 60.

54 *Spiritus in Corpore aduentitij et leuiter hærentes, spiritus rej innatos et immersos Corrumpunt et ad euolandum citant et præparant et propterea dissolutionem rej accelerant* (*De vijs mortis*, in Bacon 1996, 290, translated on 291).

55 *Corpora illa a primâ auulsione statim ad aerem exponantur i debent in frigido, ac in umbrâ, ubi calor Solis interiores spiritus sollicitare non possit, aer autem possit leuiores illos allicere et extrahere* (*De vijs mortis*, in Bacon 1996, 290, translated on 291). Graham Rees translates "allicere" as "draw out" and "extrahere" as "extract".

He seems to have fruit in mind for this particular piece of advice – it is difficult to imagine what else would be 'picked' – but the work also features similarly practical comments about animals: "in the parts of animals exercise greatly promotes longevity because it makes the members and the whole mass of the body compact and cramps the spirit. Cold has the same effect, as does bare skin, or work and living in the open and in a tolerable climate. Moreover, all astringent medicines contract the parts and restrain the spirit. And therefore they conserve and strengthen them fittingly".[56] He continues with a description of the other side of the coin: "by contrast, a sedentary life together with hot weather, too much living indoors, too many clothes, delicious baths, and soft ointments relax the parts and dilate the spirits so that their dispersal is easier and, through this, the dissolution of the thing".[57]

Bacon's thoughts on the prolongation of life have been thoroughly documented by Graham Rees, Benedino Gemelli and Guido Giglioni, so there is little point in me going over the same ground here. It is, however, worth pointing out where I differ from these authors: in the motivations that they ascribe to Bacon's work in the field. In his introduction to the *De vijs mortis*, for example, Graham Rees highlights Bacon's belief that "philosophy could improve material conditions and so in part restore prelapsarian felicity", with the prolongation of life acting as "an outstanding fulfilment of a programme proposing a material soteriology for this world".[58] Benedino Gemelli writes of the same religious urge "to intervene in the course of human history and in that of the natural evolution of man, to bring mankind back to a point where it can re-acquire the longevity that it has never definitively lost".[59] Meanwhile, Guido Giglioni suggests that the extension of man's lifespan was important not only for its religious significance, but also "as a striking validation of his new metaphysics of matter".[60] While I have no intention of dismissing the idea that Bacon may have thought of his activity in such grandiose terms, it strikes me that we can also read his work in a far more intimate context: he may, at least in part, have been motivated by the experience of his own limited, languishing body.

[56] *in partibus animalium, exercitatio multu*m *valet ad diuturnitatem vitæ, quia reddit membra et vniuersa*m *corporis massam compactam et spiritu*m *cogit in angustius. Idem facit frigus, et nuditas pellis vel labor et mora sub dio et in aeris temperæ tolerabilj. Etia*m *omnia medicamenta astringentia partes condensant arctant et spiritu*m *constringunt. Itaque Quare et propriè conseruant et roborant* (*De vijs mortis*, in Bacon 1996, 298, translated on 299).

[57] *contra vita sedentaria et eo temperie calida et mora sub tecto et vestis calid nimia, et balnea voluptaria, et mollia vnguenta laxant partes et dilatant spiritus, vt facilior sit dissipatio ipsoru*m, *vnd et per hoc dissolutio rej* (*De vijs mortis*, in Bacon 1996, 298, translated on 299).

[58] Rees 1996, lxv.

[59] Gemelli 2012, 156.

[60] Giglioni 2005, 129.

4.3 Experiments

Another strand of Bacon's intellectual activity in the years following his fall from grace in 1621 involved conducting (or at least supervising) a large number of experiments. Mentions of experimentation are dotted throughout the *Novum organum* and the natural histories, but they are particularly prominent in a voluminous collection of natural historical information, the *Sylva sylvarum*, that was published just after his death, as well as in bundles of his working notes, some of which were published in 1679, with others to be found in manuscript in the British Library.[61] Bacon's experimental practice has been the focus of some considerable attention of late, in part as the result of an article by Lisa Jardine, in which she explored his interest in 'experientia literata',[62] a phrase that is now widely deployed, although Bacon also – indeed, more commonly – uses the term 'experimentum'.[63]

Much of this recent work has sought to establish, in a general sense, Bacon's conception of the value of the knowledge that can be gained from experimental sources.[64] It is, however, also worth starting from a closer perspective and trying to identify what he hoped to gain from carrying out a particular experiment. As a jumping-off point, let us take this passage from the *Sylva sylvarum*:

> An orange, lemon, and apple, wrapt in a linen cloth, being buried for a fortnight's space four foot deep within the earth, though it were in a moist place and a rainy time, yet came forth no ways mouldy or rotten, but were become a little harder than they were; otherwise fresh in their colour; but their juice somewhat flatted. But with the burial of a fortnight more they became putrefied[65]

What is the purpose of this experiment? The most obvious motivation is pragmatic and horticultural; indeed, in the *Sylva sylvarum*, Bacon continues: "It were a profitable experiment to preserve oranges, lemons, and pomegranates, till summer; for then their price will be mightily increased".[66] There may, however, be another incentive lying behind his actions. In the *Historia vitæ & mortis*, he observes: "Fruits such as pomegranates, lemons, apples, pears and the like, and flowers such as the rose and lily keep for a long time when sealed in earthenware pots".[67] "In general", he concludes, "things put underground and in cellars, or deep underwater,

---

[61] Bacon, *Sylva sylvarum*. The notes published in 1679 are known as his *Physiological Remains* and his *Medical Remains*. A partial transcription of the notes contained in British Library Additional MS 38693 is available in Rees 1981, 394–412.

[62] Jardine 1990.

[63] For Bacon's use of the terms 'experientia' and 'experimentum', see Fattori 2002.

[64] For a sense of the recent work on the subject, see Georgescu 2011; Giglioni 2013; Jalobeanu 2011; Jalobeanu 2013; Jalobeanu 2015, 227–80; and Pastorino 2011.

[65] *Sylva sylvarum*, in Bacon 1857–59, vol. 7, 467. For a study of Bacon's experiments on fruit, see Jalobeanu 2016.

[66] *Sylva sylvarum*, in Bacon 1857–59, vol. 7, 467.

[67] Fructus, *veluti* Granata, Cytria, Mala, Pyra, *& huiusmodi; etiam &* Flores, *vt* Rosa, Lilium, *in Vasis Fictilibus benè obturatis, diutiùs seruantur* (*Historia vitæ & mortis*, in Bacon 2007, 166, translated on 167).

stay fresh longer than things above ground".[68] A similar discussion about the preservation of fruit appears in the earlier *De vijs mortis*; here, Bacon comments: "it is obvious in many instances that confinement and detention of spirit in the body is the very essence of all conservation".[69] He illustrates the principle with reference to a number of examples, one of which is the fact that "fruits [...] and particularly the juicier of them (grapes, for example) keep well for a time in flour, which is of itself able to keep the air out be it in a body or collected together. Much the same in fact can be said of wax, honey and sugar, not only in respect of fruits but of meats too".[70]

This is, however, not the end of the matter; Bacon continues: "the fact that spirit kept back, which in other circumstances would have been ready to act only on the thinner parts of the thing, tears apart and destroys the more solid parts is a matter of certainty".[71] He illustrates the point: "Thus, as I have said, it is better to use flour, fine sand, or smooth powders for keeping fruit than wax or anything else that stops the spirit's transpiration dead in its tracks".[72] There is, then, a delicate balance to be struck: the inanimate spirit has to be restrained a little, but not too much. As Bacon acknowledges: "it is [...] a difficult business and it needs to be undertaken with great circumspection for the line between softening and putrefaction is a very fine one".[73]

The very presence of such an in-depth discussion in a work on the ways of death should suggest that Bacon is not simply interested in the preservation of fruit for horticultural reasons. In fact, the context soon becomes explicit:

> This is, moreover, a wholly noble enterprise in which all care must be exercised, especially seeing that living beings and the human body most of all can receive infinite remedies and helps, helps which in no way coincide with those for inanimate things, but by which we might keep the benefits of detaining the spirit and yet ward off and restrain by appropriate means of various kinds the resulting ill-effects[74]

---

[68] *Generalitèr quæ sub* Terrâ, *& in* Cellis *subterraneis, aut in profundo Aquarum reponuntur, virorum suum diutiùs tuentur, quam quæ supra Terram* (*Historia vitæ & mortis*, in Bacon 2007, 168, translated on 169). Graham Rees translates "in profundo Aquarum" as "in the depths of the waters".

[69] *manifestum est in plurimis, quod cohibitio et detentio spiritus in Corpore omnis conseruationis præcipuum fundamentum* est (*De vijs mortis*, in Bacon 1996 306, translated on 307).

[70] *fructus* [...] *atqué ex ijs etiam humidiores (ut vuæ) bené diú seruantur in farinâ, quæ et ipsa potis est aerem (in corpore siué congregatum) excludere. Etiam de Cerâ, et Melle, et Saccharo similia affirmantur, non solum quoád fructus, sed etiam quoád carnes* (*De vijs mortis*, in Bacon 1996, 306, translated on 307).

[71] *quod spiritus retentus (qui alias solummodó in tenuiores rei partes acturus fuisset) partes solidiores digerat et conficiat; res certa* (*De vijs mortis*, in Bacon 1996, 308, translated on 309).

[72] *Itaque melius (ut diximus) adhibetur farina, aut arena subtilis et pulueres læuigati ad fructus seruandos* [...] *quam Cera, aut aliquid aliud quod transpirationem spiritus magis peremptorié prohibeat* (*De vijs mortis*, in Bacon 1996, 308, translated on 309).

[73] *negotium* [...] *scrupulosum, et quod magnâ cautione transigi vult. Arcta enim et angusta via est inter intenerationem et putrefactionem* (*De vijs mortis*, in Bacon 1996, 308, translated on 309).

[74] *Atque omninó res digna est in quâ omnis adhibeatur diligentia, præsertim cum viuentia et maximé Corpus humanum infinita suscipiat remedia et auxilia, quæ inanimatis nullo modo competunt, per quæ beneficium detentionis spiritus recipi possit, et maleficium quod ex eo sequitur nihilomi-*

It is difficult to know how literal a translation to expect between, on the one hand, Bacon's experiments on the preservation of fruit and, on the other, his interest in the prolongation of human life. The former certainly serve to bolster a theoretical framework in which detaining the inanimate spirit in some way is a beneficent exercise, but we have no record of an experiment in which he buried a person in a linen cloth for a fortnight. This section of the *De vijs mortis* closes with some suggestive remarks about the preservative powers of body paint,[75] but, on the whole, Bacon seems to have neglected the use of a physical barrier in favour of another means of constraining the spirit of a human subject: astringent medicines.

We have already encountered one deployment of astringent medicines, Bacon's maceration of rhubarb. His *Medical Remains* contain lengthy lists of astringent substances, including agarwood (lignum aloes), cloves, myrrh, pomegranates and rosemary, as well as rhubarb "received by infusion" and, in the list of purgative astringents, rhubarb, "especially that which is toasted against the fire".[76] In his *Life of Bacon*, alongside the maceration of rhubarb, William Rawley mentions Bacon's habitual consumption of saltpetre (nitre), "whereof he took in the quantity of about three grains in thin warm broth every morning, for thirty years together next before his death".[77] The *Medical Remains* feature Bacon's suggestion: "To continue my broth with nitre; but to interchange it every other two days, with the juice of pomegranates expressed, with a little cloves, and rind of citron".[78] Elsewhere in the collection is a recipe "Against the waste of the body by heat":

> Take sweet pomegranates, and strain them lightly, not pressing the kernel, into a glass; where put some little of the peel of a citron, and two or three cloves, and three grains of ambergrease, and a pretty deal of fine sugar. It is to be drunk every morning while pomegranates last[79]

In the *Historia vitæ & mortis*, Bacon is particularly expressive about the astringent qualities of another item included in the *Medical Remains*, opium; he writes, for example: "The effectiveness of opium in condensing the spirits is quite remarkable, since maybe three grains of it in short order so curdle the spirits that they do not bounce back, but are extinguished and rendered immobile".[80] A little later, he notes:

> Now the opiates can be taken not just by mouth but also by fumigation, but it should be in such a way as not to stimulate the expulsive faculty too much, or to draw out the humours, but only to work on the spirits within the brain for a short interval. Thus, a morning

nus per accommoda diuersi generis prohiberi et reprimi (*De vijs mortis*, in Bacon 1996, 310, translated on 311).

[75] *De vijs mortis*, in Bacon 1996, 310.

[76] Bacon 1857–59, vol. 3, 830 and 831.

[77] Rawley 1857–59, 16–17.

[78] Bacon 1857–59, vol. 3, 832.

[79] *Ibid.*, 829.

[80] *Efficacia* Opij *ad* Condensationem Spirituum *admodùm insignis est; cùm tria fortassè Grana eius,* Spiritus *paulò post, ita coagulent, vt non redeant, sed extinguantur, & reddantur immobiles* (*Historia vitæ & mortis*, in Bacon 2007, 246, translated on 247). Graham Rees does not translate the phrase "sed extinguantur".

> suffumigation, taken through the mouth and nostrils, with tobacco mixed together with lignum aloes, dried rosemary leaves and a pinch of myrrh, would be useful[81]

These comments serve, quite clearly, as evidence of Bacon's habit of self-medicating, but I also think they serve as evidence of a process of self-experimentation: his life seems to have been characterised by a constant urge to try out different substances, in different quantities, in different ways and in different combinations. In fact, his lifelong practice of experimental self-medication may have concluded in a moment of nasty irony: in their biography, Lisa Jardine and Alan Stewart suggest that Bacon's death was the result not, as the traditional story has it, of a chill caught while testing the preservative effects of cold on a chicken stuffed with snow, but of complications following an overdose of opiates.[82] Jardine and Stewart's narrative strikes me as the more plausible of the two accounts. It is, however, worth noting that in both cases his death is possessed of a twofold irony that would, I suspect, have appealed to the early modern (somewhat dark) sense of humour: not only was Francis Bacon the famous experimentalist undone by an experiment; that experiment was conducted as part of an investigation into the prolongation of life.

4.4 Conclusion

In recent years, historians have been laying new emphasis on the place of medicine and, in particular, on the place of medical empiricism in the development of experimental science.[83] After reviewing the material discussed here, it is difficult not to try somehow to slot Francis Bacon into the story. Of course, he was not medically trained and, although he may have shared thoughts and advice with like-minded individuals,[84] his roster of patients was, so far as we know, limited to himself. This last point, however, may be a significant one. Hands-on medical practitioners of the seventeenth century, as Harold J. Cook writes, often argued strongly that "they were far better and far more knowing" than the university-educated physicians, because

---

[81] *Possit etiam esse Acceptio* Opiatorum, *non tantùm per* Os, *sed etiam per* Fumos; *Sed talis esse debet, vt non moueat nimis Facultatem Expulsiuam, aut eliciat Humores; sed tantùm breui morâ, operetur super* Spiritus *intra cerebrum; Itq*ue Suffumigatio matutina, *per Os, & Nares excepta, cum* Tobacco, *admisto* ligno Aloes, *& Foliis Siccis* Rorismarini, *& parum* Myrrhæ, *vtilis foret* (*Historia vitæ & mortis*, in Bacon 2007, 250, translated on 251). Graham Rees translates "eliciat" as "draw down" and "*per Os, & Nares excepta*" as "swallowed or sniffed".

[82] Jardine and Stewart 1998, 502–11.

[83] Harold J. Cook has argued powerfully for the place of medicine in the shift known as the 'scientific revolution'; see Cook 2011. For a sense of the place of medicine in the development of experimental science, see Dear 2006, 111–5. For recent work on medical empiricism, see, for example, Pomata 2011 and Stolberg 2013.

[84] Lisa Jardine and Alan Stewart's analysis of Bacon's final letter, which was sent to the Earl of Arundel and contains a description of the circumstances in which he fell ill, assumes the recipient's awareness of his self-medicating practices; see Jardine and Stewart 1998, 504–5.

"they had learned by doing".[85] When it comes to his self-administered doses of astringent medicine, Francis Bacon had also learned by doing. In his case, however, the efficacy of his interventions was established not by close observation of how his patients responded, but by close observation of how he himself felt as a result.

The burgeoning scholarly interest in Bacon's conception of the importance of experimental knowledge has, not wrongly, served to paint him as a thinker preoccupied with the limits of everyday sensory experience and concerned with improving the connection between the mind and the world.[86] In a recent essay, Peter Anstey even places him alongside Robert Boyle and Robert Hooke as a theorist of what he refers to as "a representative early modern philosophy of experiment".[87] Francis Bacon certainly thought a great deal about the knowledge that could be gained from experiments, but, as this essay has shown, he also had a more personal, more intimate experience of experimentation. When his body brought forth sulphurous and fetid black bile and he was filled with feelings of lightness and peace, the conclusion that his purgative pills had worked was based not on philosophical reflection but on physical sensation. Given the emphasis throughout his work on the limits of the senses, it perhaps sounds counterintuitive to say so, but the material presented here suggests that his conviction about the value of experimental knowledge may have owed more to what he *felt* than it did to what he thought.

Bibliography

Allen, Gemma, ed. 2014. *The letters of lady Anne Bacon*, *Camden Fifth Series* 44. Cambridge: Cambridge University Press.

Anstey, Peter R. 2014. Philosophy of experiment in early Modern England: The case of Bacon, Boyle and Hooke. *Early Science and Medicine* 19 (2): 103–132.

Bacon, Francis. 1857–59. *The works of Francis Bacon*, 7 vols., ed. James Spedding, Robert Leslie Ellis and Douglas Denon Heath. London: Longman et al.

———. 1861–74. *The letters and the life of Francis Bacon*, 7 vols, ed. James Spedding. London: Longman et al.

———. 1996. *The Oxford Francis Bacon 6. Philosophical studies, c.1611-c.1619*, ed. Graham Rees. Oxford: Clarendon Press.

———. 2000a. *The Oxford Francis Bacon 4. The Advancement of Learning*, ed. Michael Kiernan. Oxford: Clarendon Press.

———. 2000b. *The Oxford Francis Bacon 13. The Instauratio magna: Last Writings*, ed. Graham Rees. Oxford: Clarendon Press.

———. 2004. *The Oxford Francis Bacon 11. The Instauratio magna Part II: Novum organum and Associated Texts*, ed. Graham Rees, with Maria Wakely. Oxford: Clarendon Press.

———. 2007. *The Oxford Francis Bacon 12. The Instauratio Magna Part III: Historia naturalis et experimentalis: Historia ventorum and Historia vitæ & mortis*, ed. Graham Rees, with Maria Wakely. Oxford: Clarendon Press.

[85] Cook 2010, 11. For an overview of the tensions between physicians and other medical professionals, see Cook 1994.

[86] For recent work on Bacon's conception of experiments, see above, n. 64.

[87] Anstey 2014, 104.

Cook, Harold J. 1994. Good advice and little medicine: The professional authority of Early Modern English Physicians. *Journal of British Studies* 33 (1): 1–31.

———. 2010. Victories for empiricism, failures for theory: Medicine and science in the seventeenth century. In *The body as object and instrument of knowledge: Embodied empiricism and early modern science*, ed. Charles T. Wolfe and Ofer Gal, 9–32. Dordrecht: Springer.

———. 2011. The history of medicine and the scientific revolution. *Isis* 102 (1): 102–108.

Dear, Peter. 2006. The meanings of experience. In *The Cambridge history of science. Volume 3: Early modern science*, ed. Katherine Park and Lorraine Daston, 108–131. Cambridge: Cambridge University Press.

Fattori, Marta. 2002. *Experientia-Experimentum*: Un confronto tra il corpus latino e inglese di Francis Bacon. In *Experientia: X colloquio internazionale*, ed. Marco Veneziani, 243–258. Florence: Leo S. Olschk.

Gemelli, Benedino. 2012. The history of life and death: A 'spiritual' history from invisible matter to prolongation of life. *Early Science and Medicine* 17 (1–2): 134–157.

Georgescu, Laura. 2011. A new form of knowledge: *Experientia literata. Societate şi Politică* 5 (2): 104–120.

Giglioni, Guido. 2005. The hidden life of matter: Techniques for prolonging life in the writings of Francis Bacon. In *Francis Bacon and the refiguring of early modern thought: Essays to commemorate the advancement of learning (1605–2005)*, ed. Julie Robin Solomon and Catherine Gimelli Martin, 129–144. Aldershot: Ashgate.

———. 2013. Learning to read nature: Francis Bacon's notion of experiential literacy (*Experientia literata*). *Early Science and Medicine* 18 (4–5): 405–434.

Gowland, Angus. 2011. *The worlds of renaissance melancholy: Robert Burton in context*. Cambridge: Cambridge University Press.

Jalobeanu, Dana. 2010. The philosophy of Francis Bacon's natural history: A research programme. *Studii de Ştiinţă şi Cultură* 23: 18–36.

———. 2011. Core experiments, natural histories and the art of *Experientia literata*: The meanings of Baconian experimentation. *Societate şi Politică* 5 (2): 88–103.

———. 2013. Learning from experiment: Classification, concept formation and modelling in Francis Bacon's experimental philosophy. *Revue roumaine de philosophie* 57 (1): 75–94.

———. 2015. *The art of experimental natural history: Francis Bacon in context*. Bucharest: Zeta Books.

———. 2016. Bacon's apples: A case study in Baconian experimentation. In *Francis Bacon on motion and power*, ed. Guido Giglioni, James A.T. Lancaster, Sorana Corneanu and Dana Jalobeanu, 83–113. Berlin: Springer.

Jardine, Lisa. 1990. *Experientia literata* or *Novum organum*? The Dilemma of Bacon's scientific method. In *Francis Bacon's legacy of texts*, ed. William A. Sessions, 47–67. New York: AMS Press, 1990. First published 1985. *Experientia literata* ou *Novum organum*? Le dilemme de la méthode scientifique de Bacon. In *Francis Bacon: Science et méthode*, ed. Michel Malherbe and Jean-Marie Pousseur, 135–57. Paris: J. Vrin.

Jardine, Lisa, and Alan Stewart. 1998. *Hostage to fortune: The troubled life of Francis Bacon*. London: Victor Gollancz.

Manzo, Silvia. 2009. Probability, certainty and facts in Francis Bacon's natural histories: A double attitude towards Skepticism. In *Skepticism in the modern age: Building on the work of Richard Popkin*, ed. José R. Maia, Gianni Paganini Neto and John Christian Laursen, 123–137. Leiden: Brill.

Marshall, Joseph, and Sean Kelsey. 'Weldon, Sir Anthony'. In The Oxford Dictionary of National Biography, accessed online via www.oxforddnb.com.

Martin, Julian. 1992. *Francis Bacon, the state, and the reform of natural philosophy*. Cambridge: Cambridge University Press.

Pastorino, Cesare. 2011. Weighing experience: Experimental histories and Francis Bacon's quantitative program. *Early Science and Medicine* 16 (6): 542–570.

Pomata, Gianna. 2011. "A word of the Empirics": The ancient concept of observation and its recovery in early modern medicine. *Annals of Science* 68 (1): 1–25.

Rawley, William. 1857–59. The life of Francis Bacon. In *The works of Francis Bacon*, ed. James Spedding, Robert Leslie Ellis and Douglas Denon Heath, 7 vols. Vol. 1, 3–18. London: Longman et al.

Rees, Graham. 1977. Matter theory: A unifying factor in Bacon's natural philosophy? *Ambix* 24 (2): 110–125.

———. 1981. An unpublished manuscript by Francis Bacon: *Sylva sylvarum* drafts and other working notes. *Annals of Science* 38 (4): 377–412.

———. 1984a. Francis Bacon and *Spiritus vitalis*. In *Spiritus: IV colloquio internazionale*, ed. M. Fattori and M. Bianchi, 265–281. Rome: Edizioni dell'Ateneo.

———, assisted by Christopher Upton. 1984b. Commentary. In *Francis Bacon's natural philosophy: A new source. A transcription of Manuscript Hardwick 72A with Translation and Commentary, BSHS Monographs,* 51–78. Chalfont St. Giles: British Society for the History of Science.

———. 1996. Introduction. In *The Oxford Francis Bacon 6. Philosophical studies c. 1611–1619*, xvii–cxvi. Oxford: Clarendon Press.

Serjeantson, Richard. 2002. Natural knowledge in the *New Atlantis*. In *Francis Bacon's New Atlantis: New interdisciplinary essays*, ed. Bronwen Price, 82–105. Manchester: Manchester University Press, 2002.

Stewart, Alan. 2003. *The Cradle King: A life of James VI and I*. London: Chatto & Windus.

Stolberg, Michael. 2013. Empiricism in sixteenth-century medical practice: The notebooks of Georg Handsch. *Early Science and Medicine* 18 (6): 487–516.

Vine, Angus. 2011. Commercial commonplacing: Francis Bacon, the waste-book, and the ledger. In *Manuscript miscellanies 1450–1700*, ed. Richard Beadle et al., 197–218. London: The British Library, 2011.

Weldon, Anthony. 1651. *The court and character of King James. Whereunto is now added the court of King Charles: Continued unto the beginning of these unhappy times*. London: J. Collins.

Chapter 5
Material Thoughts: Robert Hooke's Theory of Memory

Felicity Henderson

Robert Hooke is known for his contributions in a wide range of scientific fields, from geology to physics, but his contribution to the science of the mind, or soul, as it would more commonly have been expressed in the early-modern period, is less well-known. This is partly due to its brevity. The main source is a short lecture printed in Hooke's *Posthumous Works* by his friend Richard Waller, where the text concludes a series of lectures on light delivered in the early 1680s. Waller himself pointed out the seeming incongruity of the theme: in a short introductory note he wrote that Hooke had left the subject of light 'in some sense imperfect' in order to veer off into a new topic; and at the end of the lecture he commented 'Here our Author leaves off, nor as I can find, ever reassumed this Subject'.[1] Indeed, the subject is unique amongst Hooke's published work: it outlines a theory of the human mind, concentrating in particular on memory, the perception of time, and the action of the soul in forming new ideas.[2] It is not, however, an aberration in

I would like to thank Sachiko Kusukawa, Will Poole, Laura Sangha, Peggy Yoon, and in particular Gideon Manning, for their helpful comments on an earlier draft of this chapter. I would also like to take this opportunity to record my gratitude to Lisa Jardine for her friendship, enthusiasm and support for all things Hooke-related.

[1] Hooke 1705, 138, 148. I argue here that the lecture can be integrated into Hooke's wider research, but we should bear in mind that it may not represent his final thoughts on the topic. We no longer have the manuscript copy of the lecture but we can assume that like much of the material in the *Posthumous Works*, Hooke probably did not intend it to be printed in its current form.

[2] David Oldroyd drew attention to similarities in subject-matter between the lecture on memory and the contents of a manuscript attributed to Hooke now in Trinity College Library; however, the authorship of the manuscript text is not certain and its argument is rather different from the printed lecture. See Oldroyd 1980.

F. Henderson (✉)
University of Exeter, Exeter, UK
e-mail: F.C.Henderson@exeter.ac.uk

G. Manning (ed.), *Testimonies: States of Mind and States of the Body in the Early Modern Period*, Archimedes 57,
https://doi.org/10.1007/978-3-030-39375-5_5

Hooke's otherwise more mechanical pursuits. Rather, the lecture on memory is a point where his theories on a number of topics intersect: it draws on his work in microscopy, astronomy and physics, and it offers a new perspective on his work on geophysics. We can connect the lecture with Hooke's ongoing discussion of scientific method, particularly his 'philosophical algebra', which he believed would provide a structure and guide for philosophical thought in the same way that algebra did for geometry and mathematics. Perhaps most significantly, Hooke's theory of mind provides a strong argument for the efficacy of the Baconian method of natural philosophy, and thus for the experimental practice of the early Fellows of the Royal Society. Radically different from the models suggested by earlier writers and his contemporaries, Hooke's concept of memory and the formation of ideas can be seen to emerge directly from his own experimental practices. In essence, it is a sketch of how he conceived of his own mental processes and as such it gives us an unusual insight into the mind of an early-modern polymath – or at least, into what Hooke may have seen as the ideal mind of such an individual.[3]

While relatively scant attention has been paid to the lecture on memory, probably because of its seemingly anomalous place in Hooke's collected works, it has been treated in isolation by historians of psychology, who have discussed it in the context of early theories of the mind.[4] An important departure from this tradition was provided by John Henry, who linked for the first time Hooke's theory of memory with his experimental work in physics. Henry's argument, that Hooke's works on memory and physics demonstrated his attachment to occult notions of sympathy and antipathy, was later challenged by Mark E. Ehrlich in another useful article discussing Hooke's relevant physical theories.[5] Nick Wilding has discussed Hooke's theory of memory particularly in terms of its connection with the written records of the Royal Society, while Lotte Mulligan has concentrated on its significance in relation to Hooke's personal record-keeping and in particular his diary.[6] Rhodri Lewis has linked it with the interest among Royal Society Fellows in the classical *ars memoriae*.[7] The most detailed analysis is that of Jamie Kassler, who has provided a rich context for Hooke's work on the mind, and has made perceptive links with his other work on, for example, the vibration of springs. However her conclusion that Hooke envisaged the brain as a musical instrument, and more specifically a bell, is not entirely convincing.[8] While all these critics make useful

[3] One important part of Hooke's practice that the lecture on memory does not fully address is the role of written documents, which I discuss briefly below; see also Richard Yeo, 'Before Memex: Robert Hooke, John Locke, and Vannevar Bush on external memory', *Science in Context* 20 (2007), 21–47.

[4] See for example Singer 1976; Oldroyd 1980; Macintosh 1983; Draaisma 2000, esp. ch. 3, and 2006, 111–121; and Hintzman 2003.

[5] Henry 1989; Ehrlich 1995.

[6] Wilding 2006; Mulligan 1992.

[7] Lewis 2009.

[8] Kassler 1995. The most obvious objection to Kassler's theory is that if Hooke had conceived of the mind as a bell, he would have said so explicitly; it is also difficult to reconcile Hooke's insistence on the materiality of memories and ideas with the bell/musical instrument analogy.

comments about Hooke's work on memory, none fully explores its relevance to his scientific method. In this chapter I will introduce the main points of the lecture on memory and discuss its significance in terms of Hooke's understanding of the human body and the acquisition of knowledge about the natural world.

5.1 Time and the Human Moment

Hooke's lecture on memory begins with time. 'And here, since it is a general Maxim in the Schools, that *Nihil est in Intellectu, quod non fuit prius in Sensu*, I would query by what Sense it is we come to be informed of Time; for all the Information we have from the Senses are momentary, and only last during the Impressions made by the Object.'[9] The answer, for Hooke, is memory: the memory is an organ for perceiving time.[10] But it was these momentary impressions made by objects that prompted Hooke's lecture on memory at such a seemingly odd moment, during a course of lectures on light. Hooke had been explaining how light moves through space. He argued that light travels in pulses through a transparent medium, and that it must travel outwards '*in orbem*', that is, as an expanding sphere. The problem was how rays of light from many points could each 'act upon the inlightned Body with the same Power, as if it acted singly'. Hooke's solution lay in the infinite divisibility of matter, time and space: 'as there are infinite Parts in the least sensible Part, and infinite Moments in the least sensible Moment; so there may be infinite Parts of Space in the least sensible Space to be moved'. He argued that

> there are in every *sensible* Point of Matter a sufficient number of distinct Particles to convey every one of those Motions distinct, without interfering one with another: For as there may be many Millions of Motions communicated to a *sensible* Point, so there may be as many Millions of distinct Particles to receive each of them distinctly.[11]

Thus it would be possible for millions of motions to balance themselves out in a way that was impossible for the human eye to discern. It is not only matter that is composed of millions of distinct particles: every 'sensible Moment' of time is also composed of 'infinite Instants'.[12] This means that 'within a moment of time that is

---

Kassler herself pointed out that her model only partially fits Hooke's description (149).

[9] 'There is nothing in the understanding that was not first in sense' (from Thomas Aquinas, *De Veritate*, q. 2 a. 3 arg. 19); Hooke 1705, 139. Paul. F. Cranefield, in his note on the history of this phrase, pointed out that Thomas Willis had used it, drawing, he suggested, on Gassendi (Cranefield 1970, 77–80). It may well be from Willis that Hooke took his inspiration. Descartes attributed the ability to link together momentary impressions into a perceived whole to the imagination (Clarke 2003, 85–6).

[10] Human perception of time had been discussed since Aristotle: see Čapek 1987, 595–608); Edwards 2013.

[11] Hooke 1705, 134.

[12] Hooke was not alone in speculating on the divisibility of time. Geoffrey Gorham has compared Descartes's statement that a lifespan could be divided into 'countless parts' with the ideas of

sensible to a Man, the same sensible Point may be moved many Millions of ways successively'. Hooke explained this with the examples of the vibrations of a taut string, and a glowing coal being moved quickly in the dark: in both cases, the human eye perceives the object 'as if it were in all parts of its space', whereas moved more slowly, the string or coal can be perceived as a discrete object being moved.[13]

The idea depends on the concept of the 'sensible moment': the smallest length of time in which humans can perceive discrete occurrences. Hooke argued that 'the sensible Moments of Creatures are somewhat proportion'd to their Bulk', and that those creatures that seem short-lived in respect to humans probably experience as many distinct moments of time in the course of their short lives as men do in their long ones. As evidence for this he cited his observation of the beating of a fly's wings, which to humans seem merely a blur, but to the fly probably have at least three distinct moments (a lower, middle and upper point). This idea of the 'human moment' clearly intrigued Hooke. He read a further lecture to the Royal Society on 17 May 1682, which was described in the official minutes as 'about local motions, sensible and insensible times, and celerity'.[14] In his own memoranda, though, Hooke wrote 'Lecture of Humane moments', suggesting that this was the key point for him.[15] For Hooke the microscopist, whose observations assured him that unimaginably small spaces really exist, the infinitely divisible nature of the human moment perhaps raised the intriguing possibility of a device that might allow philosophers to see into the gaps between instants. Just as our human perception of gold saw a uniformly solid metal when in fact, Hooke argued, it is mixed with aether, so human perception of time is merely a facade behind which innumerable other things are happening insensibly.[16]

Thus it was his speculations about time and the human moment that led Hooke to think about the mechanism of memory. Because the information from the senses is only momentary, '[t]here is therefore yet wanting a Sense to apprehend Time'. And yet humans are conscious of time passing (at least while they are awake). The answer, for Hooke, was the organ of memory, which, he suggested, acts as a kind of internal clock by counting off sensible moments. Hooke had a model for this in the weather-clock invented by Christopher Wren in the early 1660s and improved by himself over the ensuing decades. This instrument combined a pendulum-clock with a rain-gauge, barometer, thermometer, hygroscope, and wind vane; the clock driving a rotating cylinder which unwound a paper strip onto which readings from the instruments were automatically recorded every 15 min.[17] The device thus linked

---

Gassendi and Isaac Barrow, both of whom also thought that time was composed of successive instants (Gorham 2007, 44, 50–51). Hooke's interest though was not in the theological necessity for God's continual creation but in human perception.

[13] Hooke 1705, 134. For more on Hooke's theory of motion see Ehrlich 1995, 142ff.

[14] Birch 1756–57, IV: 148.

[15] Henderson 2007, 153.

[16] Hooke 1705, 172.

[17] Nick Wilding has discussed the relationship between the weather-clock and Hooke's theory of memory (2006, 124); see also Biswas 1967, 94–104.

data with time in a newly-verifiable way. As the Royal Society's institutional records make clear, however, the readings produced by the weather-clock were not immediately intelligible. In April 1674, Hooke was asked to help the Society's assistant 'to reduce into writing some of the first papers marked by the weather-clock', and when he brought in a written description of the instrument he was asked to add to it 'the key for expounding the papers marked by the engine'.[18] Data-points and time-references were not enough to make sense of the phenomena they recorded – for that, some kind of guiding intellect was required. In his theory of the human mind, this was the task of the soul.

5.2 Memory and Reason

If memory is an organ for perceiving time, it can only operate through the interaction of the soul with the senses:

> Memory … I conceive to be nothing else but a Repository of Ideas formed partly by the Senses, but chiefly by the Soul it self: I say, partly by the Senses, because they are as it were the Collectors or Carriers of the Impressions made by Objects from without, delivering them to the Repository or Storehouse where they are to be used. Which Impressions being actual Motions … those Motions conveyed to this Repository become Powers sufficient to effect such Formations of Ideas as the Soul does guide and direct them in: For I conceive no Idea can be really formed or stored up in this Repository, without the Directive and Archi[t]ectonical Power of the Soul; and the Actions or Impressions cease and fail without the concurrent Act of the Soul, which regulates and disposes of such Powers.[19]

This act of the soul in regulating sensory impressions Hooke calls 'attention'. As the impressions are delivered to the brain the soul forms tiny particles of matter, 'some material Part of the Repository', into specific shapes with their own motions, and these are the memories that retain the sensory impressions.[20] These particles or memories are produced constantly (while the senses are engaged), and remain in the repository in the order in which they were formed. They spiral outward from the 'Center' where the soul works, and this 'Chain of Ideas' is responsible for mankind's perception of time: the distance between new ideas in the 'Center' and those further along the chain is interpreted by the soul as the passing of time, and the length of the chain corresponds directly to the length of intervening time between perceptions. 'And so', Hooke concludes, because these distances and lengths can (theoretically) be measured, 'Time comes to be apprehended as a Quantity, and so falls under the Consideration of Geometry and Mensuration.'[21] Despite his obvious

[18] Birch 1756–57, 4: 277, 283. For early attempts to record weather patterns see Daston 2015.

[19] Hooke 1705, 140.

[20] This description of the soul acting to form ideas clearly worried some members of the Royal Society, who thought it made the soul 'mechanical' – see below.

[21] Hooke 1705, 141. Hooke does not offer any evidence for this 'notion' of his concerning the memory as a repository of ideas formed by the soul, nor does he give a location for the soul in the

interest in the mathematics of the human perception of time, this comment is not directly concerned with such measurements – rather, I think Hooke meant to say that it is only because memories are formed in this chain-like progression that humans are able to apprehend time as a quantity; the fact that we do so is due to the fundamental design of our minds rather than any quality inherent in time itself.[22]

Having arrived at a model of memory that explained our perception of time to his own satisfaction, Hooke clearly felt that it behoved him to explore the ramifications of his theory in more detail. Perhaps the most unusual aspect was his conception of memories and ideas (he uses these words somewhat interchangeably) as physical particles.

> These Ideas I will suppose to be material and bulky, that is, to be certain Bodies of determinate bigness, and impregnated with determinate Motions, and to be in themselves distinct; and therefore that no two of them can be in the same space, but that they are actually different and separate from one another; and as they have their distinct Figures, so have they each of them their distinct Qualifications of Motions and Constitutions.[23]

The ideas are made from matter in the brain which is capable of receiving sensory impressions: 'one peculiar Kind for the Impressions of Sight', another for impressions of hearing, touch, and so on. In case his auditors baulked at the idea of matter with such unusual properties Hooke reminded them of the '*Bononian* stone', a phosphorescent substance which was capable of absorbing light and re-emitting it (albeit for a short time).[24] Hooke posits that the body might be able to make materials with similar capacities for storing sensory impressions – not just of light, but also of feeling, taste, smell, and sound. To give further weight to his argument for this 'adapted Matter' produced in the brain, Hooke explains that it acts in a similar way to the mechanisms of the senses themselves, which are suited to individual species and gather information specific to that species.

---

brain, although he claims that he could say more on the matter (Ibid.). The tone of the lecture is clearly speculative, although, as this chapter argues, the theory emerges from work that Hooke has done previously.

[22] Following the second reading of his lecture on memory at the Royal Society meeting on 28 June 1682, Hooke showed the Fellows 'a very easy way of plainly demonstrating the problem of Archimedes, whereby he proved, that the tangent of his spiral at the point of one revolution from the center will cut a ray produced from the center at right angles to the ray from the touched point; so that the part between the point of intersection and the centre shall be equal to the circumference of the circle, that passes through the point of the spiral, where the said tangent touches it' (Birch 1756–57, 4: 154). Given that in Hooke's model memories emerge in a spiral from the centre, it seems likely that this discussion was linked (at least in his own mind) with the mathematics of our perception of time.

[23] Hooke 1707, 142.

[24] Draaisma has discussed Hooke's metaphor of the Bononian (Bolognian) stone (2006, 112–4). Hooke may also have had in mind Frederick Slare's recent demonstrations of the properties of liquid and solid phosphorus, which he showed to the Royal Society several times during the early 1680s. Slare wrote that both types of phosphorus were made 'out of substances taken from a Humane Body', and they occasioned at the Royal Society a discussion of whether this might prove the existence of a *flammula vitæ*, a suggestion with which Hooke disagreed (Slare 1681; Birch 1756–57, 4: 125).

> As for instance, the Smell being caused by a subtil and curious Exhalation from the odoriferous Body imbibed by the Air, the Olfactory Nerves are prepared with an aerial Body fit to dissolve or imbibe that Substance in the same manner as the Air does from the odoriferous Body; which aerial Body, by means of the Olfactory Nerve, having an immediate Intercourse and Passage to the Brain, does immediately convey it thither: And according to the nature of this aerial or spirituous Substance with which the Olfactory Nerve is furnished, so does it dissolve or imbibe this or that Exhalation out of the Air.

Thus dogs are able to smell the scents of animals or flesh, imperceptible to men; and Hooke suggests that men are able to smell flowers, plants and fruits which dogs may not smell at all. This understanding of animal bodies as radically different from each other (and humans) and adapted for different functions recurs throughout Hooke's work; the point here is that Hooke extends this argument to the operations of the brain.[25] Like the body, the brain is a biochemical machine where the 'Chymistry of Nature' manipulates substances in 'the wonderful Elaboratory of the Animal Body'.[26]

The soul impresses the memory-particles with form and motion; but of these two attributes, Hooke places far greater emphasis on motion than form. He does at one point call ideas 'little Images, which bear the Stamp, Seal or Mould' given them by the soul when they were formed, but again this description stresses the three-dimensional physicality of an object impressed with a shape, rather than a two-dimensional image. Ideas or memories are set in motion by the power of the soul 'radiating' onto them like the sun's rays. Both form and motion decay over time, as memories are jostled against one another in the repository, but as long as they remain in motion memory-particles also have the power of acting on the soul, 'to excite it to Attention':

> … and by this means also whenever any Idea is created and impregnated with Motions or Qualifications similar to those of other Ideas placed at some distance in the Repository, the concurrent Impressions or Re-actions of those similar Ideas upon the Soul at that time do make the fainter to be the more notable, and so excite the Soul to Attention or Radiation that way also; … And this I take to be that Impression which we are sensible of, when we say, This brings to my Mind, or This puts me in mind, or this makes me remember, *&c*.[27]

The soul is aware of most thoughts in the sphere of the repository (except those that have degraded or had other thoughts interposed between them and the centre), and by radiating its attention onto them it can renew their initial force, at the same time creating new ideas at the centre with similar properties.[28] These new ideas have a 'Sympathetick Agreement' with the old memories which makes both particles stronger in the same way that a musical string can make another move in unison, making the resulting note louder.

---

[25] Hooke 1705, 142; compare Hooke's description of the action of smelling in Hooke 1665, sig. c1^r. For the different sensory perceptions of animals see Hooke 1705, 8–9; 1665, 178–9.

[26] Hooke 1705, 141.

[27] Ibid., 145.

[28] Hooke describes the soul's various actions as 'Attention', 'Apprehending, Remembring and Reasoning', and 'Thinking' (1705, 140, 141, 145), but he never mentions the imagination, which would more commonly have been the part of the mind that forms new ideas.

This harmony of new and old thoughts not only acts to reinvigorate older memories, but is also a crucial part of thinking, which Hooke says is partly memory and partly the creation of new ideas. New ideas are formed when the soul compares

> the Re-actions from several Ideas placed here and there in the Repository, and [is] sensible of the Harmony or Discord of them one with another, which does produce an Idea wherein all those various Respects are in some means united and impressed upon one and the same Idea. This is an Idea of greater Perfection, and according to the Attention of the Soul in being sensible of more and more variety of former Ideas, and the Regularity and Order of its proceeding in that Action, and the more steddy and distinct manner in the Course and Progress of it, so is the Idea more compleat, as well as more compounded: And this I conceive to be that Action of the Soul which is commonly called Reasoning; and the Conclusion is the new Impression made upon the Idea informing from the comparison of other Ideas[29]

An idea 'of greater perfection' arises from the ability of the soul to link together or compare a number of different ideas with similar properties, ideas that are resonating in harmony with one another. Clearly those with a better stocked and well-ordered repository will be able to locate a greater number of harmonious ideas and therefore will have a better chance of creating new, more perfect ideas. And since the raw materials from these ideas or memories comes from sensory perceptions, those with more and wider experiences of the physical world will have a better stocked repository.[30]

The 'storehouse' metaphor for memory was of course not new, but Hooke's choice of the word 'repository' in this context was probably not accidental.[31] Hooke was the curator of the Society's Repository, or museum of artefacts, and as Matthew Hunter has pointed out Hooke's role as curator is mirrored here in the soul's activities in the repository of memory.[32] (We might also note that the diverse group of people who donated objects for the Repository are thus loosely equated with the senses.) What was the significance of this metaphorical connection with the Repository? We might recall the early claim by Thomas Sprat that the Society's collection held 'the greatest part of all the several kinds of things, that are scatter'd throughout the *Universe*' – an ambition towards comprehension that was perhaps mirrored, for Hooke, by a dream of universal knowledge.[33] Nehemiah Grew's catalogue of the Repository, first published in 1681 (that is, around the same time that Hooke's lecture was written) showed how the artefacts could be arranged into schemes based on their underlying similarities. Such classification schemes, exemplified by John Wilkins's tables in his *Essay towards a Real Character and*

---

[29] Ibid., 146.

[30] This, and a number of Hooke's other ideas about perception and memory, are reminiscent of those put forward by Locke in *An essay concerning human understanding* (London, 1690), but this does not necessarily indicate influence. On the two philosophers' conceptions of memory see Yeo 2007.

[31] For example, John Beale mentioned the Tradescant collection in this context (John Beale to Robert Boyle (29 September 1663), in Boyle 2001, 2: 137. Locke uses both 'Store-house' and 'Repository' for the memory (Locke 1975, II.x.2 (p. 150)).

[32] Hunter 2013, 178–83.

[33] Sprat 1667, 251.

Philosophical Language, were not only based on perceived similarities but were intended to reveal new connections. Primarily though, the Repository functioned as a kind of external institutional memory for the Royal Society, holding items presented and discussed at meetings – items that could be recovered and reused or re-investigated if necessary.

The final section of Hooke's lecture is devoted to his explanation of the soul and its actions. That this was a significant, and controversial, aspect of the lecture is demonstrated by the comments of his immediate audience and his posthumous editor. After its second reading at a Royal Society meeting, we are told that some Fellows objected 'that this discourse seemed to tend to prove the soul mechanical … Mr Hooke answered, that no such thing was hinted, or in the least intended in it'. Waller, in a note at the conclusion of the printed text, admitted that 'possibly some Persons may imagine that the foregoing Explication of these abstruse Actings of the Soul is too mechanical, and tends to the making the Soul a material Being'; both here and in his preface to the *Posthumous Works* he found it necessary to reassert the soul's spiritual nature, on Hooke's behalf.[34] Hooke himself calls the soul 'an Incorporeal Being', and 'the *Primum movens*, the self-moving Principle'. Despite its incorporeal nature, though, the soul makes use of corporeal organs to perform its actions, 'and without them cannot effect what it wills', be this seeing, hearing or remembering. Hooke admits that he 'cannot conceive' how this happens, but he does offer an extended metaphor, or similitude, to describe the workings of the soul: 'the best Similitude for that purpose … is the Sun in the great World'.[35] The sun's influence, or light, extends everywhere. Hooke argues that if the sun were conscious it would be able to sense all the bodies its rays of light touches, because the end of the ray would meet with resistance. Not only this, but the sun would know how far away the impeding body was, because of the strength of the resistance to its rays; and it would know how large the object was because of the number of rays being resisted. In the same way, the soul is sensible of all the ideas in the repository, and their relative positions. Furthermore, the sun exerts 'an attractive Power' over all the bodies in its sphere, and thus influences their motions.

> And hence the Bodies so placed, as they have each of them peculiar Properties, Shapes and Motions of their own, so have they also particular Influences, Radiations, Excitations and Regulations communicated to them from the Sun, which gives them not only their regulated Motions and Positions, but also a kind of new Being or Activity, by which they become visible and sensible to the rest of the World, which would otherwise be dark and insensible, and vagrant here and there uncertainly in the *Expansum* of the Universe.

---

[34] Birch 1756–57, IV: 154; Hooke, 1705, 148 (italics reversed), sig. A2v. I think 'mechanical' here has the meaning intended by Boyle when he discussed the 'Mechanisms of the … Human Body', that is, unconscious or reflex actions. Boyle used the example of the pupil's dilation and constriction: 'these various Motions in the Eye are produc'd by mere Mechanism, without the Direction, or so much as Knowledg or Perception, of the Rational Soul' (Boyle, *A free enquiry into the vulgarly receiv'd notion of nature* (1686), in Boyle 1999–2000, *10:* 538, 541).

[35] Hooke 1705, 140, 146.

Thus the soul, Hooke says, 'forms to it self a Microcosm ... in which it radiates, and is sensible of every thing contain'd therein, in the same manner as the Sun in the Macrocosm.'[36]

In his musing on the soul, Hooke returned again to the question that prompted the lecture in the first place, the propagation of light. This suggests that the image of sun as soul was fundamental to Hooke's theory of mind rather than simply a helpful way of suggesting a possible mechanism of interaction between incorporeal and material elements of the system. As John Henry has argued, the similitude with its references to microcosm and macrocosm seems to draw Hooke back into an occult world of sympathies and active principles. However if we assume that it was his contemplation of the sun's place in the universe that influenced Hooke to draw up the memory-system that he did, I think we need to interrogate the metaphor in a way that concentrates on its mechanical aspect. Hooke himself encourages us to do this when he describes the physical relationship between the sun and the objects its rays light upon: the 'Impediment ... must be reciprocally proportionate to the Distances of the impeding Object'.[37] This is not the language of occult beliefs. What Hooke is saying is that the fundamental laws that govern the universe, some of which he had described in print many years earlier, also govern the workings of the human mind.[38] Despite not being absolutely sure about the physics of attraction and repulsion, he can see no reason why those same rules would not apply to every part of the natural world.[39]

The potential for action at a distance means that Hooke's theory of mind is not confined to the individual body. The final paragraph of his essay conjectures that the soul's sphere of radiation extends beyond the repository, throughout the body, and even possibly outside the body.

> Now though by what I have been saying, I have endeavour'd to shew that the Soul has by its Radiation a more than ordinary and commanding Power over all the Ideas placed within the Repository; yet I would not be understood so to limit its Sphere of Radiation, as not to suppose that it may have a much bigger Sphere of influencing Power, and thereby may extend it, not only to all and every Point of the Body inlivened and preserved by it; but possibly it may extend even out of the Body, and that to some considerable Distance, and thereby not only influence other Bodies, but be influenc'd by them also.

As instances that suggest such an influence Hooke goes on to note 'the Sensibleness of others Ideas, *Lupus in Fabula*, *Fascination*, &c'.[40] Although Hooke had clearly

---

[36] Ibid., 147.

[37] Ibid.

[38] Hooke 1674, esp. pp. 27–8.

[39] It may seem unusual for Hooke to have moved from physics to the soul in this way. Thomas Sprat famously announced that the Royal Society in its early years had avoided enquiries into only 'two Subjects, *God*, and the *Soul*', but he also suggested that long study of the human body would enable there to be 'very neer ghesses made, even at the more *exalted*, and *immediate* Actions of the *Soul*; and that too, without destroying its *Spiritual* and *Immortal* Being.' (Sprat 1667, 83). For a discussion of the science of the mind as part of early-modern natural science see Hatfield 1995.

[40] Hooke 1705, 147. 'The wolf in the story', a Latin phrase roughly equivalent to the English 'Speak of the devil': presumably Hooke was suggesting that the mind is sometimes aware of

not given this much thought, and finishes with the rather anti-climactic 'of which possibly some other time', it is an intriguing addition to the theory. The claim that the soul could interact with ideas in other repositories suggests that everyone's memories are made of the same material, and reinforces the notion that the mind and its contents conform to physical laws that are uniform and work the same way in each individual.

In my discussion here I want to suggest some new ways in which Hooke's theory of mind can be linked with his other research, and perhaps more importantly, with his methodology. As other scholars have pointed out, Hooke's description of the workings of the mind draws heavily on his physics. It hinges on the idea of particles vibrating in harmony, an essential component of his theory of congruity, the theory he used initially to explain why certain substances mix together and others do not, but which eventually became fundamental to his theory of the universe.[41] He posited that all particles in the universe are in motion. Those with similar motions 'vibrate together in a kind of Harmony or unison' and are therefore attracted to one another. Dissimilar particles 'though they may be both mov'd, yet are their vibrations so different, and so untun'd, as 'twere to each other, that they cross and jar against each other, and consquently cannot agree together, but fly back from each other to their similar particles.' The particles in the memory behave in a similar fashion, but instead of creating new chemical substances, harmonious memory particles create new ideas.[42]

The physicality of Hooke's memory-particles is striking, and sets his theory apart from most other contemporary theories of memory and the mind. Hooke's conception of memory gives primacy to the physical – that is, sensory experiences and material manifestations of them in the mind. Ideas are 'things' in a very material sense, and they are prompted by physical things in the external world: memories of sorrow, joy, fear or love are never mentioned and it is difficult to imagine how they would be formed into particles for insertion into the repository. The mind Hooke describes is that of an experimental philosopher, and the ideas it contains seem to be wholly material. In other memory-systems of the time, thoughts and ideas were commonly described as images or patterns. Descartes, for example, thought sensory impressions made patterns in the material of the brain.[43] Locke denied the materiality of ideas altogether, although he used the metaphor of

---

someone approaching before their arrival. Bacon described fascination as 'the power and act of Imagination, intensiue vpon other bodies, than the bodie of the Imaginant' (Bacon 2000, 105).

[41] On Hooke's theory of congruity see Ehrlich 1995.

[42] Here Locke seems to have differed from Hooke, arguing that 'Wit' lies in quickly putting together ideas which had 'resemblance or congruity', whereas 'Judgment' lies in carefully distinguishing between ideas, 'to avoid being misled by Similitude' (Locke 1975, II.xi.2 (p. 156)).

[43] See Desmond Clarke 2003, esp. ch. 2; and Danziger 2012, esp. pp. 43–4. Danziger and Draaisma give very useful historical overviews of the metaphors used for memory, which was regularly represented as a storehouse, building, or form of writing – but these I think should be regarded separately from the speculations of Hooke, Descartes and others on the physical workings of the mind (Danziger 2012, esp. pp. 24–58; Draaisma 2000).

'pictures' freqently.[44] Responding in part to Descartes, Kenelm Digby posited a version of the extramission theory that coincided, of all the contemporary theories of sense and memory, most closely with Hooke's ideas. Digby argued that there is 'a perpetuall fluxe of litle partes or atomes out of all sensible bodies', and that these atoms 'gett in at the dores of our bodies', mingle with the animal spirits, and are conveyed to the brain where they wait in 'some vacant cell' until called for by the memory.[45] It is entirely possible that Hooke was influenced by Digby's theory – with the important difference that while Digby's sensations were atoms produced by external things, Hooke's memories were formed by the body's own material in response to sensations. As mentioned above, Hooke himself used the word 'images' to explain his concept of material ideas at one point in his lecture: 'certain Ideas, which for the present I will call little Images, which bear the Stamp, Seal or Mould according to which the Soul formed it'.[46] This is, I think, merely a convenient and well-understood metaphor used by Hooke in this instance for ease of comprehension: his main argument here is about the action of the soul. Elsewhere his ideas are always referred to as particles. Unlike an image stamped in wax, which waits passively to be viewed or read, Hooke's idea-particles have motion and a sphere of influence of their own. Because it is grounded so firmly in his microscopical researches, I would argue that the shift in Hooke's theory to memories as particles opens the way for a wholly new, biochemical approach to the mind-body relationship.

5.3 Unnatural Man: The Relationship Between the Senses and Knowledge

In order to understand Hooke's mind we need to understand his body. For Hooke, the human body was an instrument that was simultaneously powerful and fallible. Among his writings there are regular comments about the limitations of the body – specifically, limiting its usefulness for the pursuit of natural philosophy.[47] Certain knowledge is gained via the senses, but the human senses are dull and require assistance. This is because they were not designed for acquiring philosophical knowledge, they were designed (like those of all animals) to support the basic functions of life – the sight to identify dangers and opportunities, the smell and taste for assessing the wholesomeness or otherwise of food – rather than for anything more

[44] Locke 1975, II.x.2 (p. 150).

[45] Digby 1644, 278, 284. There are other similarities: Digby's memories are stored sequentially (or at least in proximity to those formed at the same time); he uses the metaphor of sympathetic vibration (to explain stronger memories of more frequent sensations); and he explains the fact of forgetting things by physical damage to memories over time (Ibid., 284–287).

[46] Hooke 1705, 144.

[47] See, for example ibid., 6–15.

sophisticated.[48] Crucially, the senses shape and limit what each species of animal knows about the world: Hooke suggested that things may taste or smell different to humans than to other animals, and argued that the 'Constitution of Air, which we call dark, is not so to Cats and Owls'.[49] Thus 'our Apprehensions of things seem to be appropriated to our Species: And that if there were another Species of Intelligent Creatures in the World, they might have quite another kind of Apprehension of the same thing, and neither perhaps such as they ought to be, and each of them adapted to the peculiar Structure of that Animal Body in which the Sensation is made.'[50] Thus the ideas that each animal can produce are a direct result of its own particular fabric, and the ideas of humans about the natural world are not necessarily any more accurate than those of any other animal. It is entirely consistent for Hooke not to give primacy to the human view of the world. He argued that 'We ought to conceive of things as they are part of, and Actors or Patients in the Universe, and not only as they have this or that peculiar Relation or Influence on our own Senses or selves.'[51] His goal was to understand the system of the universe as a whole – seeing how things were interconnected, influencing each other entirely apart from human concerns. At the same time man is still part of this system, and is governed by its laws and logic, which may itself colour the 'perception' of the system.

If ideas are shaped by the body, then changes to the body will result in changes to the understanding. Much of Hooke's discussion on this theme centres on external aids for the senses, but he does seem to accept the possibility of a more widespread shift in human perception. As Nick Wilding and others have pointed out, Hooke's conception of the defective body is linked with the idea of the Biblical Fall (though not in a particularly detailed way): he wrote optimistically that 'as at first, mankind fell by tasting of the forbidden Tree of Knowledge, so we, their Posterity, may be in part restor'd by the same way, not only by beholding and contemplating, but by tasting too those fruits of Natural knowledge, that were never yet forbidden'.[52] Hooke went through the motions of looking back to a pre-lapsarian time when bodies were more perfect, but he remained convinced that men would be able to 'recover some degree of those former perfections' by 'rectifying the operations of the Sense'. Through this process of enlightenment, 'all our command over things is to be establisht'. He began the Preface to *Micrographia* with a bold statement about man's difference from animals: men are able 'not only … to behold the works of Nature, or barely to sustein our lives by them', but they have 'the power of considering, comparing, altering, assisting, and improving them to various uses' – or, in other words, to subject them to scientific scrutiny and then turn the results to economic

---

[48] Ibid., 8.

[49] Ibid., 9.

[50] Ibid., 8. These views were not unique to Hooke but were held, with some variations, by Locke, Descartes and others: see Shapiro 2010, 265–285.

[51] Hooke 1705, 9. This of course marks a shift from the providential view of the universe, in which everything had been created specifically for mankind's use.

[52] Wilding 2006; Hooke 1665, sig. b[2]$^{r-v}$.

and social benefit.[53] This is the goal of all Hooke's speculations about body, mind and reason – the desire to increase knowledge about the natural world with the aim of improving mankind's use of it. Hooke suggests that this process, if applied to the human senses, will produce a virtuous circle: by using the senses to understand the works of nature, we will better be able to identify the ways in which the senses are deficient, and find ways to repair them. And despite his Biblical language here, in fact we find that this natural knowledge would improve mankind, intellectually and physically, to such an extent that philosophical man would exceed common men 'as much, almost, as they do Brutes or Ideots'.[54] In essence, a new race of super-humans will emerge whose powers of cogitation will be almost unrecognisable to current men.

I would like to suggest that this rhetoric of improvement has more in common with Hooke's theories of metamorphosis, species change over long periods of time resulting in almost unrecognisable new forms, than it does with the narrative of the Fall and its consequences for human knowledge. Hooke wrote a long series of lectures on geology that were grouped together and printed by Waller under the general heading 'A Discourse of Earthquakes', but were written and refined over a forty year period from the mid-1660s to near the end of Hooke's life. Like much of his work, the theory is underpinned by microscopic observations and was probably sparked by his research for *Micrographia*.[55] The lectures outline Hooke's theory of geological change: that is, the Earth's surface has undergone constant change through the action primarily of earthquakes (in which category Hooke includes volcanoes), but also of the movement of water, sand and air; and that this violent change explains the creation of fossils and their current locations.[56] Fossils were once living creatures, and this means that there were once many other species that no longer exist, and probably also that there are now many new kinds that did not exist in the beginning.[57] Not only do species blink in and out of existence, but they may also change shape:

> we will grant also a supposition that several *Species* may really not have been created of the very Shapes they now are of, but that they have changed in great part their Shape, as well as dwindled and degenerated into a dwarfish Progeny; that this may have been so considerable,

---

[53] Hooke 1665, sig. a1^{r}. This passage recalls Boyle's comments in *Some considerations touching the usefulness of experimental natural philosophy* (1663) (Boyle 1999–2000, 3: 212).

[54] Hooke 1705, 3; a more forceful version of his statement in the Preface to *Micrographia*, that human nature is capable of such improvement 'as to make some Men excel others in their Observations, and Deductions, almost as much as they do Beasts' (1665, sig. a1^{r}).

[55] For example, he argues that fossils are formed when the 'Pores' in vegetable and animal substances are filled with 'some petrifying liquid Substance' (Hooke 1705, 290); that the insides of plants, animals and minerals are specific to their species or type and conform to the outsides (ibid., 291–2); and discusses the 'regular Figures of Salts' and other crystalline substances (ibid., 294).

[56] This is a very truncated version of Hooke's theory; for further discussion see Poole 2010; Drake 1996.

[57] Hooke 1705 291, 327–8, 435–6, 450 *et passim.*

> as that if we could have seen both together, we should not have judged them of the same Species.[58]

This is a narrative of decline, prompted by the fact that living nautilus species are much smaller than the ammonite fossils found in England; Hooke goes on to argue that all things, terrestrial and celestial, tend towards decay and dissolution but that this is no argument against the 'Omnipotence, Providence and Wisdom of the Creator'.[59] Humankind is not mentioned here, but Hooke's commitment to rationality would not have allowed him to exempt humans from the kinds of changes undergone by other species even if his instinct for self-preservation prevented him from making such a statement publicly. But did change necessarily mean decline? Elsewhere he speculates that there may have been 'a preceding learned Age' when all the arts had been brought to perfection before being lost in one of the regular 'Devastations' suffered by the Earth.[60] I would argue that Hooke's faith in the scientific method, expressed so forcefully elsewhere, led him to look forward to another learned age of the future, when men will not just equal but outperform Adam. What Hooke was suggesting was that science, with its 'artificial Instruments and methods', would eventually re-make mankind.[61]

Obviously, though, this was not going to happen overnight. In the meantime Hooke had some ideas about how the fruits of natural knowledge might more readily be tasted – through improvement of the dull human senses. Microscopes and telescopes, obviously, could extend the power of seeing, perhaps infinitely: why not build a thousand-foot telescope, if it were possible to grind large enough glasses for it? More powerful microscopes might also allow men to perceive motions that were normally too fast or slow for human sight: Hooke uses the examples of 'the Legs of Mites' and the hand of a watch, and looks forward to a time when he might be able to see plants growing.[62] Other senses might be improved in other ways.

> 'Tis not improbable also, but that our taste may be very much improv'd, either by preparing our tast for the Body, as, after eating bitter things, Wine, or other Vinous liquors, are more sensibly tasted; or else by preparing Bodies for our tast; as the dissolving of Metals with acid Liquors, make them tastable, which were before altogether insipid; thus Lead becomes sweeter then Sugar, and Silver more bitter then Gall, Copper and Iron of most loathsome tasts. And indeed the business of this sense being to discover the presence of dissolved Bodies in Liquors put on the Tongue, or in general to discover that a fluid body has some

---

[58] Ibid., 435.

[59] Ibid.

[60] Ibid., 328. This speculation is linked with Hooke's continuing interest in Chinese culture and chronology, which stretched back far earlier than that of Western chroniclers. See Hooke 1686, 63–78, esp. pp. 64–5; Poole, 2012, 135–153.

[61] Hooke 1665, sig. a1^{r}. Hooke's fellow-microscopist Henry Power was equally enthusiastic about the potential of the new philosophy, placing philosophers 'in a rank Specifically different from the rest of groveling Humanity', and arguing that even before the Fall the constitution of Adam's body was the same as modern men's, and thus he could never have seen what modern men now could through telescopes and microscopes (Power 1664, 192 and sig. [A4]$^{r-v}$). On the significance of Adam to the new philosophers see Picciotto 2010.

[62] Hooke 1705, 15–16.

> solid body dissolv'd in it, and what they are; whatever contrivance makes this discovery improves this sense… . 'Tis not improbable also, but there may be multitudes of other wayes of discovering the parts dissolv'd, or dissoluble in liquors; and what is this discovery but a kind of secundary tasting.[63]

Here we see both a clear (and rather horrifying) picture of what Hooke was willing to subject his body to in the name of scientific discovery, and also the way that he categorised what we would consider to be straightforwardly chemical processes as 'a kind of secundary tasting' – that is, an external procedure that allowed the same kind of perception as the sense of taste. Because the purpose of taste is to 'discover the presence of dissolved Bodies in Liquors put on the Tongue, or in general to discover that a fluid body has some solid body dissolv'd in it, and what they are', then 'whatever contrivance makes this discovery improves this sense'. The body of the experimenter still has primacy: the processes of nature are still only understandable via sense perception. But if the body is a complex machine, then it is a continually adaptable one: other bits of machinery or other mechanical processes such as the dissolution of tasteless solid substances can be fitted onto the body like a prosthetic limb.

Why insist on the chemical process described here as a kind of 'secundary tasting'? Partly, of course, to prove Hooke's point about the possibility of improving the senses – but partly, I think, so that the data produced can be compared like for like. Hooke returns several times to the requirement for as many comparable instances of phenomena as possible, recommending that his readers try to compress data into a small space so that it can all be seen in one view. Perhaps categorising all instances of solids dissolved in liquids under the heading of taste allowed for easier categorisation and comparison: properly prepared, lead was 'sweeter then Sugar', silver 'more bitter then Gall'. What were the properties that produced these tastes, and were they comparable in substances as seemingly dissimilar as lead and sugar, silver and gall? If the sense of taste was what converted material substances in the natural world into material ideas inside the mind, then this 'secundary tasting' provided an added body of food for thought.[64] Viewed in this way, the human body becomes a site of transmutation, where the matter of the external world is processed into memory-particles for storage in the brain.

It should be clear from this description that Hooke's theory of memory and the mind fits well with his ideas about the philosophical body. Both emphasise movement and interaction both within the bodily systems and between them and the material world, and in both there is a rhetoric of improvement consistent with Hooke's overall positive outlook for science. Hooke had a whole-body approach to science – he gave each of the senses their own sphere of activity, under the direction of the faculty of reason. In fact the continual and continued activities of the senses are vital in Hooke's conception of the project of knowing. Philosophy should:

---

[63] Hooke 1705, sig. c2^{r-v}.

[64] It is noticable that the 'business' of taste here is entirely scientific rather than pleasurable – this is not, I think, because Hooke saw the body solely as a knowledge-making machine, but rather that his purpose here is to discuss methodology.

> begin with the Hands and Eyes, and ... proceed on through the Memory, to be continued by the Reason; nor is it to stop there, but to come about to the Hands and Eyes again, and so, by a continual passage round from one Faculty to another, it is to be maintained in life and strength, as much as the body of man is by the circulation of the blood through the several parts of the body, the Arms, the Fat, the Lungs, the Heart, and the Head.[65]

Hooke argues that the experimental philosophy exceeds the 'Philosophy of discourse and disputation' because the latter ignores 'the Sense and Memory'. Here Hooke is referring to his own particular theory of memory as the record of physical sensations and objects. Discourse and disputation were of course also founded on memory, but memory of words and commonplaces rather than memory of things, and therefore, in Hooke's view, not the right sort of memory for true knowledge.[66]

Hooke's understanding of the senses and the body is influenced by his microscopy, and by his mechanical view of the universe. He sees the human mind, and the human body, as part of a much larger world system: laws applicable to (in the example described above) chemical processes also apply to bodily processes, and the same procedures lead to the same outcomes. This is a fundamental requirement for Hooke's philosophy. It was a commonplace among writers on memory that 'sensible things' were easier to store and more firmly retained in the memory than 'insensible' notions; Hooke provides a mechanism to explain why this should be so.[67]

5.4 Memory and Philosophical Practice

What might Hooke's theory of memory tell us about his own conception of his philosophical practice? We know he distrusted his own memory, and experimented with substances he hoped would improve its performance. Such medicinal preparations were a fairly general desiderata for natural philosophers in the period, as witnessed by Robert Boyle's inclusion of 'Potent Druggs to alter or Exalt Imagination, Waking, memory, and other functions' in an early list of scientific projects.[68] The materiality of Hooke's memory-particles immediately suggests the potential for alteration or improvement of the memory through the use of chemical substances, although this is not something he hints at in his lecture.

One aid for memory Hooke does discuss is writing. He uses the example of compilations of medical case-histories:

> If a Physitian be ... accounted the more able in his Faculty, because he has had long experience and practice, the remembrance of which, though perhaps very imperfect, does regulate all his after actions: What ought to be thought of that man, that has not only a perfect

[65] Hooke 1665, sig. b2^r.

[66] For contemporary theories about memory see Yates 1966.

[67] Willis 1661, sig. [A5]^r.

[68] Royal Society Archives RB/1/8/30. See Mulligan 1992, 53 for Hooke's experiments with memory-enhancing substances, and on his use of medical preparations more widely see Jardine 2003, 181–193 and her contribution to this volume.

> register of his own experience, but is grown old with the experience of many hundreds of years, and many thousands of men.[69]

Hooke's language here is instructive. As always, 'experience and practice' has primacy: it is this that gives the physician authority, and it is the memories of his experiences that regulate his action. The power of writing, though, is literally transformational. The collected case histories not only assist memories of personal experiences, but they combine to create a kind of augmented reality. The reader/physician 'is grown old' with the experiences of many thousands of others: the implication is one of bodily change, that somehow the reader becomes a palimpsest, taking on the combined personae of his antecedents. When we remember the temporal dimension to Hooke's theory of memory we can see exactly why this reader can be said to have 'grown old': his spiral of memory particles is now much longer than it normally would be for a man of his age.

Written records, then, are as much a way of controlling time as information. It was a desideratum of Hooke's to have a large amount of data presented in one view: he returns to this idea in theory and practice several times, in his own diary (which is initially set out with a month per page, with columns of weather observations alongside daily memoranda); in his various tables or lists representing classification schemes; and in his instructions for making a record of the weather, in which he specifically requires a month's weather to be presented on a single page for ease of comparison.[70] In fact the practice of Hooke's diary-keeping, and the making of classification schemes, had contemporary analogues in the writings of those teaching the *ars memoriae*, and these can be linked with Hooke's theory of memory in interesting ways. Hooke's habitual, almost obsessive listing of the names of friends in his daily diary entries has long puzzled scholars, since it is difficult to see how the bare lists of names could convey any meaning a week or a month later. However John Willis in his *Mnemonica* had this recommendation for men of affairs: 'set down your chiefest business of [the] day, and also the names of such persons as you have conversed with about any serious affair, either at home or abroad.' Although this may not seem particularly useful at the time, he maintained that it may eventually come in handy:

> Suppose that after some revolution of time, three months, a year or more, question arise about the very day whereon such or such a thing hapned, an exact knowledg whereof will be very profitable; the certain day you have forgotten, but well remember it was on the same day your sheep were shorn, or the day after such friends dined with you; this being considered, your Almanack will exhibit the particular day. Moreover by the mens names with whom you spake that day, haply you may learn many things most necessary to be known in the present cause.[71]

[69] Hooke 1665, sig. d[1]^r. Hooke may have been thinking here of Hippocrates's famous aphorism, rendered in Latin as 'Ars longa, vita brevis' ('art is long, life is short'). On medical case-histories in the period see Pomata 2010.

[70] Lorraine Daston has characterised this power as 'super-vision' (Daston 2015).

[71] Willis 1661, 2.

Hooke was possibly following similar reasoning when he noted people's names day after day: they were not merely items of information in themselves, but prompts for other memories. Again, personal experience is key (not, in Hooke's case, sheep-shearing day – but certainly the experience of dining with friends). And again, a cluster of memories is made to resonate together, provoking an idea about the particular significance of some business in which several people had been involved.

Contemporary writers on the art of memory also discussed lists that function in a similar way to the kinds of classification schemes I have already mentioned. Hooke's Royal Society colleague John Beale explained to Robert Boyle how heads and subdivisions of matters could aid the memory, using the example of a man going to a fair with '50 or 100 businesses'. Remembering these will be easy

> if he first reduceth these hunderds to soe many heades of Uses, … & subdivides every use with the remarke of concernement & the losse that oblivion may cause, This will give fullest security for the maine, & may bee branched from fewe heades into such cleare Order, as cannot be a burden.[72]

This division of matter into heads was not only an aid to memory, but could actively assist in finding out new knowledge. John Willis recommended using a set of interrogative heads to recall aspects of a business to mind; but moreover, he argued that: 'in handling any subject, many things may escape our scrutiny, which a careful repetition of these Verses, and prudent application, may prompt the mind'. He gave the example of 'a learned Counsellor', consulted about a Case in which it was important that all the 'material business' was known: if these 'Interrogations' were applied then the relevant information would follow.[73]

It seems to me that these cases are examples of different ways in which Hooke might have understood memory – either housed in the body or the external memory of records – as a source of new philosophical knowledge. Hooke certainly believed that such knowledge was almost inevitable, given large enough datasets properly arranged. Returning to the compilation of medical case-histories presenting hundreds of years' worth of experience he wrote:

> What may not be expected from the rational or deductive Faculty that is furnisht with such Materials, and those so readily adapted, and rang'd for use, that in a moment, as 'twere, thousands of Instances, serving for the illustration, determination, or invention, of almost any inquiry, may be represented even to the sight? How neer the nature of Axioms must all those Propositions be which are examin'd before so many Witnesses? And how difficult will it be for any, though never so subtil an error in Philosophy, to scape from being discover'd …?[74]

Although this passage gives primacy to sight and the medium of the written word, the 'materials' ranged for use seem to operate in a similar fashion to the mind's own memories, with the 'architectonic' soul combining both written and remembered instances in the act of reasoning.

---

[72] John Beale to Robert Boyle (29 September 1663), in Boyle 2001, 135–136.

[73] Willis 1661, 33, 37.

[74] Hooke 1665, sig. d[1]^r.

Thus written materials such as these perform a more significant function than simply recording information: they serve not just for 'illustration' or 'determination' but also for 'invention'. The passage quoted above is reminiscent of Hooke's claims for his fugitive 'Philosophical Algebra', to which he refers several times but never fully explains.[75] In order to advance knowledge, Hooke argued, the intellect required a 'Method or Engine' to assist it in the same way that instruments and experiments assisted the senses. Using the analogy of the 'Art of *Algebra* in Geometry', Hooke maintained that his 'Method' would provide 'a Guide to regulate [the intellect's] Actions, so as that it shall not be able to act amiss'.[76] This early emphasis on the notion that, given 'fit Materials to work on' and a 'Method or Engine' to guide it, the intellect would *perforce* come to a correct conclusion, has an interesting resonance with his later theory of ideas. As we have seen, in Hooke's mind it is the universal physical laws of the universe that cause memory particles to vibrate in harmony and produce ideas of greater perfection – here the 'Engine' guiding the intellect towards the truth is no longer a man-made 'Method' but God's universal law.

5.5 Conclusion

The two common themes running through all these examples are copiousness and comparison. Thus, I want to conclude by proposing a link between Hooke's theory of memory and the methods promoted by Francis Bacon. In its simplest form, Bacon's method of advancing knowledge required the collection of large amounts of data and the ability to compare it. The Royal Society's practices were still at this point essentially Baconian (insofar as we can characterise the Royal Society as having a set of agreed institutional practices); more importantly for my purposes, I think we can see that Robert Hooke's practices were also essentially Baconian. Indeed, Hooke approvingly cited Bacon as having begun to provide the kind of 'Engine' to assist the mind that he himself was proposing with his Philosophical Algebra.[77] What is interesting about Hooke's theory of memory, I would argue, is that it provides a theoretical, mechanical, explanation for why the Baconian method of science works. The key is Hooke's theory of congruity, which explains how the relevant ideas or memories coalesce into what Hooke calls 'an idea of greater perfection' – that is, a piece of knowedge or a theory about the natural world that must, through the conditions of its formation, be correct. The production of abundant memories or ideas, through sensory experience or, as we have just seen, through reading written records, is fundamental to this process. Hooke emphasises the need

[75] The most detailed discussion appears in 'A General Scheme or Idea of the Present State of Natural Philosophy', probably composed in 1666 but only published by Waller in the *Posthumous Works* (Hooke 1705, 1–65). See Hesse 1966, 67–83.

[76] Hooke 1705, 6.

[77] Ibid., 7.

for a wide personal experience of natural and man-made artefacts and processes. On an institutional level, however, the practices of the Royal Society also aid this process of gathering experiences or data. Although scholars have paid a good deal of attention to the Society's information-gathering practices in terms of correspondence, record-keeping and publications, I would argue that in this context the Fellows' weekly meetings must be given more serious consideration than has previously been the case. Anyone who has read the Society's journal books, recording the minutes of their meetings, must immediately be struck by Hooke's comment about memories recalled by other memory-particles vibrating in harmony, prompting statements such as 'This brings to my Mind, or This puts me in mind, or this makes me remember, etc.' These are precisely the kinds of free associations that Fellows made during their weekly discussions of topics that ranged over the whole of the natural world. In Hooke's conception of the production of knowledge, though, these were not merely random additions to the conversation, they were potentially part of a useful new idea. Hooke's suggestion that the soul's attention can extend beyond the boundaries of the individual skull to sense the ideas of others is also relevant here. While not precisely the hive-mind beloved of modern science-fiction writers, Hooke seems to have imagined the Royal Society's group discussions as more than the sum of their parts. By merely coming together in the same room, Fellows could be contributing to a new 'idea of greater perfection', or one that was almost axiomatic.

Hooke's understanding of the natural world was fundamentally based on the identification of what I'd like to call congruencies. (Here I'm speaking with reference to his work in *Micrographia* and his interest in instruments rather than his physics, although they cannot be completely dissociated.) Readers of *Micrographia* will be struck by Hooke's use of metaphor to explain the processes and structures he observes. It seems to me that these metaphors are not simply an explanatory tool for readers, or decorative, but that they serve an important purpose for Hooke himself in trying to understand how things worked. For example, consider Hooke's observation of the many hemispheres of a fly's eye:

> in living Flies, I have observ'd, that when any small mote or dust … chances to light upon any part of these knobs, as it is sure to stick firmly to it and not fall, though through the *Microscope* it appears like a large stone or stick (which one would admire, especially since it is no ways probable that there is any wet or glutinous matter upon these *Hemispheres* …) so the Fly presently makes use of his two fore-feet in stead of eye-lids, with which, as with two Brooms or Brushes, they being all bestuck with Bristles, he often sweeps or brushes off what ever hinders the prospect of any of his *Hemispheres*, and then, to free his leggs from that dirt, he rubs them one against another, the pointed Bristles … of which looking both one way, the rubbing of them to and fro one against another, does cleanse them in the same manner as I have observ'd those that Card Wool, to cleanse their Cards, by placing their Cards, so as the teeth of both look the same way, and then rubbing them one against another.[78]

[78] Hooke 1665, 176.

Here Hooke uses a number of metaphors, or associations, to make some observations about the fly's body. The underlying assumption is that nature acts uniformly: materials that look the same (tiny motes of dust, and a stick or stone) should behave in the same way (not sticking to a smooth dry convex surface). Similarly, the same operations (in this case, rubbing bristly objects against one another to remove unwanted material caught in the bristles) produce the same results.

I think Hooke not only thought that this procedure of making associations, or finding congruencies, was a useful way to make theories about the operations of the natural world, but that it would also point the way to new possibilities where the same operation could be used in a different context – perhaps in a mechanical device, since all these operations were mechanical anyway. This instance also brings us back to the senses and the primacy of sensory perceptions. Hooke's observation of people (probably women) doing the common and mundane task of carding wool and cleaning their cards, gave him an instance to draw upon when engaged on the seemingly unrelated task of microscopic observation. In all his coffeehouse conversations, meetings with booksellers and builders and tradesmen across London, Hooke was constantly collecting and storing memories, in an unending spiral illuminated by his soul.

Hooke's theory of memory and ideas was not (merely) an odd deviation into metaphysics prompted by his speculations on the nature of light. As I have argued, the lecture on memory connected several ongoing themes in Hooke's research, overlapping with his scientific theories about time, light and gravity, and also resonating with his speculations about the physical history of the Earth. It provides a means by which these scientific studies can be linked with Hooke's other ongoing preoccupation, the methodology of science. Hooke's interest in this has been tepidly acknowledged by scholars, largely through analysis of his writings in the Preface to *Micrographia* and in the posthumously-published 'General Scheme', but there has been little discussion of how his methodology actually affected his practice, and whether other Fellows of the Royal Society may have subscribed to similar ideas. Was Hooke attempting to formulate a template for institutional scientific research, and to reinvigorate the Society by providing a theoretical foundation for his chosen formula? The Royal Society in the 1680s continued to suffer from one of the problems that had plagued it since its foundation, lack of a clear research programme. As one of the relatively few Fellows to make experiments, Hooke was certainly well-placed to give an opinion on this.

Regardless of whether the lecture was supposed to prompt others to follow his lead or not, it certainly suggests some avenues for investigation into Hooke's own methodology as he would hardly have described a theory of knowledge that ran counter to his own experience of coming up with new ideas. His use of similitudes needs to be examined in this context; rather than simply helping to explain his ideas to others, the theory of ideas suggests that they are an active part of the knowledge-making process for Hooke. In Hooke's formulation, the nature and range of experiences (memories) one can call upon when thinking will materially affect the ideas formed. It would be very interesting to collate all of Hooke's similitudes in order to see what themes emerge: one might expect architecture and construction to play a major part, for example, but it would be instructive to know whether many similitudes were built on information that had been passed on to Hooke via other sources rather than experienced physically by him. This would be a true test of the worth of the Royal Society's meetings in terms of generating knowledge. What were all those months and years of discussions, (more or less) faithfully recorded in the Society's Journal Books, worth to Hooke? And were they worth more than his everyday experiences in the streets and coffeehouses of the City of London? If we take his comments on the primacy of the senses seriously, then perhaps it was the teeming laboratory of London that stimulated Hooke to think his greatest thoughts.

Bibliography

Bacon, Francis. 2000. In *The Oxford Francis Bacon book IV: The advancement of learning*, ed. Michael Kiernan. Oxford: Clarendon Press.

Birch, Thomas. 1756-57. *History of the royal society*, 4 vols. London: A. Miller.

Biswas, Asit K. 1967. The automatic rain-gauge of Sir Christopher Wren, F.R.S. *Notes and Records of the Royal Society* 22: 94–104.

Boyle, Robert. 1999–2000. *The works of Robert Boyle*, 14 vols., ed. Hunter, Michael and Edward B. Davis. London: Pickering & Chatto.

———. 2001. The correspondence of Robert Boyle, ed. Michael Hunter, Antonio Clericuzio and Lawrence M. Principe, 6 vols. London: Pickering & Chatto.

Čapek, Milič. 1987. The conflict between the absolutist and the relational theory of time before Newton. *Journal of the History of Ideas* 48: 595–608.

Clarke, Desmond. 2003. *Descartes's theory of mind*. Oxford: Oxford University Press.

Cranefield, Paul F. 1970. On the origin of the phrase '*Nihil est in intellectu quod non prius fuerit in sensu*'. *Journal of the History of Medicine and Allied Sciences* XXV (1): 77–80.

Danziger, Kurt. 2012. *Marking the mind: A history of memory*. Cambridge: Cambridge University Press.

Daston, Lorraine. 2015. Super-vision: weather watching and table reading in the early modern Royal Society and Académie des Sciences. In *Curiously drawn: Early modern science as a visual pursuit*, ed. Felicity Henderson, Sachiko Kusukawa, Alexander Marr, *Huntington Library Quarterly* 78: 187–215.

Digby, Kenelm. 1644. *Two treatises: In the one of which, the nature of bodies; in the other, the nature of mans soule; is looked into: In way of discovery, of the immortality of reasonable soules*. Paris: Gilles Blaizot.

Draaisma, Douwe. 2000. *Metaphors of memory: A history of ideas about the mind*, trans. Paul Vincent. Cambridge: Cambridge University Press.

———. 2006. Hooke on memory and the memory of Hooke. In *Robert Hooke: Tercentennial studies*, ed. Michael Cooper and Michael Hunter, 111–121. Aldershot: Ashgate.

Drake, Ellen Tan. 1996. *Restless genius: Robert Hooke and his earthly thoughts*. New York/Oxford: Oxford University Press.

Edwards, Michael. 2013. *Time and the science of the soul in early modern philosophy*. Leiden/Boston: Brill.

Ehrlich, Mark E. 1995. Mechanism and activity in the scientific revolution: The case of Robert Hooke. *Annals of Science* 52: 127–151.

Gorham, Geoffrey. 2007. Descartes on time and duration. *Early Science and Medicine* 12: 28–54.

Hatfield, Gary. 1995. Remaking the science of mind: Psychology as natural science. In *Inventing human science: Eighteenth-century domains*, ed. Christopher Fox, Roy Porter, and Robert Wokler, 184–231. Berkeley/Los Angeles: University of California Press.

Henderson, Felicity. 2007. Unpublished material from the memorandum book of Robert Hooke, Guildhall Library MS 1758. *Notes and Records of the Royal Society* 61: 129–175.

Henry, John. 1989. Robert Hooke, the incongruous mechanist. In *Robert Hooke: New studies*, ed. Michael Hunter and Simon Schaffer, 149–180. Woodbridge: Boydell Press.

Hesse, Mary B. 1966. Hooke's philosophical algebra. *Isis* 57: 67–83.

Hintzman, Douglas L. 2003. Robert Hooke's model of memory. *Psychonomic Bulletin & Review* 10: 3–14.

Hooke, Robert. 1665. *Micrographia*. London: Printed by Jo. Martyn and Ja. Allestry.

———. 1674. *An attempt to prove the motion of the earth*. London: Printed by T. R[oycroft] for John Martyn.

———. 1686. Some observations, and conjectures concerning the Chinese characters. Made by R. H. R. S. S. *Philosophical Transactions* 16: 63–78.

———. 1705. In *The posthumous works of Robert Hooke*, ed. Richard Waller. London: S. Smith and B Walford.

Hunter, Matthew C. 2013. *Wicked intelligence: Visual art and the science of experiment in restoration London*. Chicago/London: University of Chicago Press.

Jardine, Lisa. 2003. Hooke the man: His diary and his health. In *London's Leonardo: The life and work of Robert Hooke*, ed. Jim Bennett et al., 181–193. Oxford: Oxford University Press.

Kassler, Jamie C. 1995. *Inner music: Hobbes, Hooke and North on internal character*. London: Athlone.

Lewis, Rhodri. 2009. Hooke's two buckets: Memory, mnemotechnique and knowledge in the early Royal Society. In *Ars reminiscendi: Mind and memory in renaissance culture*, ed. Donald Beecher and Grant Williams, 339–363. Toronto: Center for Reformation and Renaissance Studies.

Locke, John. 1975 [1690]. *An essay concerning human understanding*, ed. Peter Nidditch. Oxford: Clarendon Press.

Macintosh, J.J. 1983. Perception and imagination in Descartes, Boyle and Hooke. *Canadian Journal of Philosophy* 13: 327–352.

Mulligan, Lotte. 1992. Robert Hooke's "memoranda": Memory and natural history. *Annals of Science* 49: 47–61.

Picciotto, Joanna. 2010. *Labors of innocence in early modern England*. Cambridge, MA/London: Harvard University Press.

Pomata, Gianna. 2010. Sharing cases: The *Observationes* in early modern medicine. *Early Science and Medicine* 15: 193–236.

Oldroyd, D.R. 1980. Some "philosophical scribbles" attributed to Robert Hooke. *Notes and Records of the Royal Society* 35: 17–32.

Poole, William. 2010. *The world makers: Scientists of the restoration and the search for the origins of the earth*. Witney: Peter Lang.

———. 2012. Vossius, Hooke, and the early Royal Society's use of sinology. In *The intellectual consequences of religious heterodoxy 1600-1750*, ed. John Robertson and Sarah Mortimer, 135–153. Leiden: Brill.

Power, Henry. 1664. *Experimental philosophy, in three books*. London: John Martin and James Allestry.

Shapiro, Lisa. 2010. Instrumental or immersed experience: Pleasure, pain and object perception in Locke. In *The body as object and instrument of knowledge: Embodied empiricism in early modern science*, ed. Charles T. Wolfe and Ofer Gal, 265–285. Dordrecht: Springer.

Singer, B.R. 1976. Robert Hooke on memory, association and time perception (1). *Notes and Records of the Royal Society* 31: 115–131.

Slare, Frederick. 1681. An account of several experiments made with the shining substance of the liquid and of the solid phosphorus. *Philosophical Collections* 3: 48–50.

Sprat, Thomas. 1667. *The history of the Royal Society of London for the improving of natural knowledge*. London: J. Martyn and J. Allestry.

Wilding, Nick. 2006. Graphic technologies. In *Robert Hooke: Tercentennial studies*, ed. Michael Cooper and Michael Hunter, 123–134. Aldershot: Ashgate.
Willis, John. 1661. *Mnemonica, or, the art of memory*. London: Leonard Sowersby.
Yates, Frances A. 1966. *The art of memory*. London: Routledge and Kegan Paul.
Yeo, Richard. 2007. Before Memex: Robert Hooke, John Locke, and Vannevar Bush on external memory. *Science in Context* 20: 21–47.

Chapter 6
Making Sense of Pain: Valentine Greatrakes, Henry Stubbe and Anne Conway

Sarah Hutton

Abstract This paper discusses some interconnections between explanations of pain and the experience of pain in seventeenth-century England by focusing on an unusually well-documented episode in the history of pain relief: the visit to England in 1665 by the Irish healer, Valentine Greatrakes. This is an episode which captures the complexity of the early modern medico-philosophical setting and the human experience of illness. I review the role of Henry Stubbe and discuss the role of pain in Anne Conway's philosophy to argue that her philosophical treatise reflects her acquaintance with medical theories which were in circulation at the time of the Greatrakes affair."

Keywords Anne Conway · Valentine Greatrakes · Henry Stubbe · Henry More · Pain · Early-modern medicine

In this paper I consider some of the interconnections between explanations of pain and the experience of pain in the seventeenth-century by examining an unusually well-documented episode in the history of pain relief, and the interplay of philosophical and medical theory of which it formed the focus. This was the visit to England in 1665 by the Irish healer, Valentine Greatrakes, at the behest of Lord Conway, in the hope that he might cure his wife. The events surrounding Greatrakes' very public visit to the Conway seat at Ragley Hall in Warwickshire, also happen to be a very-well documented. As a result, this episode gives us both a snapshot of the state of scientific and medical theory of the day, as well as abundant testimony to the

S. Hutton (✉)
The University of York, York, UK
e-mail: sarah.hutton@york.ac.uk

G. Manning (ed.), *Testimonies: States of Mind and States of the Body in the Early Modern Period*, Archimedes 57,
https://doi.org/10.1007/978-3-030-39375-5_6

pain and suffering which drove people to seek Greatrakes' help.[1] The episode captures the complexity of the early modern medico-philosophical setting, the competing claims of rival theories and the human experience of illness.

As the major symptom of physical indisposition, pain is a topic at the centre of medical theory and practice in all periods. The seventeenth-century was no exception. Pain was a fact of everyday life to a far greater extent than for us today since effective pain-control was limited. There is so much that modern medicine has largely consigned to oblivion—not least the chronic character of early modern illness, and the levels of suffering to be endured from both therapies and surgical interventions, as much as from illness itself in a period when analgesics were few and primitive. In the new philosophies of this period, pain figures in some key ways at the interface of body and mind. The relationship of the soul or mind to the body is not a new topic in the seventeenth century but it takes on a new importance with the shifts in philosophical frameworks that came with the decline of Aristotelianism and the rise of the 'mechanical' philosophy. That philosophers should concern themselves with a topic from the province of medicine, is not surprising given the greater inter-connectedness of the disciplines at this time, and the fact that the domain of what we now call science was a branch of philosophy. Several early modern philosophers had a strong interest in medicine: two notable examples are Descartes and Locke.[2] But it is not just in natural philosophy that seventeenth-century philosophers discuss pain: it also figures in moral philosophy (e.g. in Locke as a motivator to avoid evil) and in metaphysics. A striking example of this is Greatrakes' patient Anne Conway, in whose *Principia philosophiae antiquissimae et recentissimae* pain has a significant place. That Anne Conway should concern herself with the problem of pain has not a little to do with the fact that she was a life-long patient, suffering from an incurable condition.

There is a difference, of course, between explaining pain, the *experience* of pain, and coping with it. Whatever the speculations of learned physicians and philosophers as to the causes of pain and its purpose (if any), exhortation to patience and fortitude was the default advice for dealing with it. There is a long tradition of moralising pain in terms of God's purposes—from the suffering of the damned, to warning signs from God, to the practice of penitence, and the trials of the faithful. It seemed perfectly logical to offer spiritual counselling as a palliative for physical suffering, and remedies for pain included recommendations of spiritual reformation—a remedy which perhaps might strike us nowadays as a remedy of desperation.

[1] Hutton 2004. Greatrakes was not the only healer of this kind. Another was Matthew Coker (Nicolson and Hutton 1992, 98ff). Unlike Greatrakes, Coker laid claim to prophetic powers and the ability to cast out evil spirits. Although More's former tutor, Robert Gell, believed Coker's cures to be miraculous, More offered a naturalistic, psychosomatic explanation of Coker's gift, in terms of what he called a 'sanative contagion', an explanation which is echoed in accounts of Greatrakes' healing by Lord Conway, George Rust and even Henry Stubbe (Crocker 2005, 115–18). In a *scholium* his *Enthusiasmus Triumphatus* More recounts his telling Boyle of his explanation (More 1712, p. 51).

[2] For discussion, see Aucante 2006, Anstey 2010, Manning 2014, Milton 2001, Dewhurst 1963. For an overview of the literature on Descartes, see Manning 2007.

But this was a remedy which made sense in terms of the seventeenth-century understanding of the mutual interaction of mental and physical or what we might call 'psychosomatosis.'[3] Today, psychosomatic illness has dubious connotations and is usually regarded as a psychological rather than a medical condition. In the seventeenth-century mind-set the mutual interaction of body and mind was regarded as well-established, even if it was not fully understood. Among the topics which Sir Francis Bacon considered worthy of investigation were, 'how, and how farre the humours and temperament of the body, doe alter or worke upon the mind', and 'how the mind influences body', and 'how and how farre the Passions and apprehensions of the mind, doe alter or work upon the Body.'[4] Henry More, understood that relationship in moral terms, claiming that 'the Diseases of the Body are, for the most part, from the Vices of the Mind' while he saw a corollary between physical health and virtuous conduct: 'a purifi'd Mind goes a great way to the purging and purifying of the Body' so 'bodily Health is thus gotten and sustaind by Virtue'. In consequence, 'there is no Remedy so powerful, for such an Incumbrance, as a severe application of Virtue and Piety.'[5] As these examples suggest, the modern disciplinary division which assigns mind to philosophy and body to medicine, did not obtain in the seventeenth century.

6.1 Greatrakes, Witnesses and Interpreters

Important sources for Greatrakes' work as a healer are the witness statements which are contained in two open letters to Robert Boyle both published in 1666: the first of these, was published by the Warwickshire physician Henry Stubbe who was at that time physician to the Conways. Stubbe's *The Miraculous Conformist: or an Account of Severall Marvailous Cures Performed by the Stroaking of the Hands of Mr Valentine Greatarick* also printed the observations made by Ezekiel Foxcroft, whom Stubbe had invited to contribute. Foxcroft was a Cambridge don and chemical philosopher, whose mother, Elizabeth Foxcroft, was for a time companion to Lady Conway. The second set of witness statements is Greatrakes' own *A Brief Account of Mr Valentine Greatraks, and Divers of the Cures by him Lately Performed* (1666; reprinted Dublin 1668), which consists almost entirely of signed affidavits from a large number of people testifying to the efficacy of his cures and to his good and pious character. This abundance of witnesses included not just those who had benefited from Greatrakes's ministrations, but a substantial number of respected persons ('men not easily imposed on', as the editor of the 1723 edition called them; Greatrakes 1723, iv). Thirdly, there are the notes which Boyle made both prior to

[3] I.e. something akin to Juvenal's saying, 'mens sana corpore sano'. There is no definition of *psychosomatosis* in the *Oxford English Dictionary*. But the term has modern usage in the literature of psychology, where its associations are with mental illness.

[4] Bacon 1640, 183.

[5] More 1690, 147–8.

and during his encounter with Greatrakes, together with his letter to Stubbe in response to *The Miraculous Conformist*.[6]

Another rich source for Greatrakes and the medical and cultural context in which he operated is Anne Conway's surviving correspondence.[7] Although this contains no personal account of her treatment at Greatrakes' hands, the letters document many aspects of his visit, from Lord Conway's securing his services and to the aftermath in London. The letters also provide a context for the visit, namely Anne Conway's experience of illness, and they give us a relatively well-recorded case of extreme and prolonged personal suffering, along with a fairly detailed record of the therapeutics she underwent. What is unusual about the medical history of Anne Conway is that she was a philosopher. And being a female philosopher was particularly unusual in the seventeenth century.[8] It is also striking that pain figures in her philosophy. Thus, taken together, Anne Conway's writings exhibit an unusual confluence of personal, medical, philosophical and spiritual elements, which gives us some insight into both personal experience of illness and long term experience of pain in the seventeenth century.

Greatrakes attracted the interest of doctors, theologians and philosophers alike. These included, among the learned, Robert Boyle, Henry Stubbe, Ezekiel Foxcroft, Henry More and at least four other members of his circle (George Rust, Benjamin Whichcote, Ralph Cudworth and Joseph Glanvill). Boyle in his turn enlisted several physicians to be witnesses (William Denton, James Faireclough, and Albertus Otto Faber appear alongside his own in the testimonies printed by Greatrakes). Although not a signatory to affidavits, the physician Thomas Willis could be added to the list, since Stubbe dedicated his book on the subject to him. Although she was not 'learned' in the same sense, another witness was Boyle's sister Lady Ranelagh, in whose house Robert Boyle interviewed Greatrakes subsequently to his visit to Ragley.[9]

The apparent efficacy of Greatrakes' 'stroking' put theorists on their mettle to provide an explanation, raising issues about legitimate knowledge, worries about diabolism, and questions about miracles. There were sharp divergences among the competing explanations of Greatrakes' apparent successes.[10] There were those who saw his cures straightforwardly as miracles or as evidence of the supernatural.[11] Others explained them in terms of spirits,[12] while physicians were among those who testified to their effectiveness. Robert Boyle abstained from speculating as to the

[6] Hunter 2009.

[7] Nicolson and Hutton 1992.

[8] For the increasing attention to women's contribution to philosophy, see, inter alia Broad 2003 and the special issue of *The Monist* (2015) edited by Karen Green and Ruth Hagengruber. On Conway as a philosopher see Hutton 2004.

[9] On Katherine Jones, Lady Ranelagh, see Hutton 2004, chapter 6, and Michelle DiMeo 2015.

[10] For discussion, see Kaplan 1982; Steneck 1982; Jacob 1983; Cook 1989; Elmer 2013.

[11] Anon. 1666; Greatrakes 1723.

[12] Crocker 2005.

causes of Greatrakes' cures.[13] In short, as Anne Conway's brother, the Solicitor General Heneage Finch apparently told Greatrakes, he had caused 'the greatest faction and disturbance between clergy and laymen that anyone has seen these past 1000 years.'[14]

The depositions by witnesses of Greatrakes's activities as a healer, especially those by Stubbe and Foxcroft, offer us a glimpse of the experience of illness, the sheer desperation of the incurable and the uncured, as well as the pain and physical repulsiveness of illness. In *The Miraculous Conformist*, Stubbe supplemented his published account of Greatrakes' healing with a further set of observations supplied by Ezekiel Foxcroft, Stubbe's account spares no graphic detail in his description of Greatrakes's activities, for example

> I saw him launce a Wenne that covered the Eye of an old man; there issued out abundance of matter in smell, and consistence, and colour, resembling a rotten-Egge.[15]

Ezekiel Foxcroft is no less graphic in his recording of the afflictions of those who sought Greatrakes' help. For instance, he writes of one woman

> I never saw any Hysterical Paroxysm half so dreadful; all her Joynts seemed to be crampt, and sometimes so distorted, as if She had been upon the Wrack. ... And it was very observable, that whither soever her Pain retired, that part was (as it were) crampt, and seem'd by her Motions, Cryings out, and Shrieks, to be affected with intolerable Dolour; sometimes it was in her Throat, and then it nigh choaked her, and there ensued excessive Belchings...[16]

The descriptions supplied in the multiple affidavits in Greatrakes' book are less graphic than those of Stubbe and Foxcroft. Most commonly the affidavits collected by Greatrakes describe the cures in terms of relief from pain. Although some of those cured describe some details of their maladies, there is an emphasis on the types of condition he healed rather than physical particulars. Benjamin Whichcote, was one of those to whom Greatrakes's stroking brought relief. Although he describes the malady from which he himself suffered, he gives more extended testimony to the type of condition which Greatrakes treated:

> I have seen too in very many instances by his spittle and the touch or stroking of his hand, humours put into odde and violent fermentations, pains strangely flye before him, till he has chased them out at some of the extreme parts of the body; the Kings-evil in a few days wonderfully dryed up, knobs or kernels brought to a suppuration; tumours ripened; ulcerous sores skinn'd and amended; hard swellings in womens breasts abated; cold and senseless limbs restored to heat and life; scabs all over the body, which have been for many years, and counted incurable, deadned and dryed up; many people relieved in cases of deafness, lameness, dimness of sight; twenty several persons in Fits of the Falling-sickness, or Convulsions,

[13] Boyle 2007; Hunter 2009.

[14] Nicolson and Hutton 1992, 272. Explanations of Greatrakes's cures were complicated by the fact that he had something of a controversial background, partly because of his claim to have cured the 'king's evil', but also because of his radical associations (he was a friend of the regicide Robert Phaire in whose regiment he served during the Civil War, and was acquainted with Lodowick Muggleton (Elmer 2013)).

[15] Stubbe 1666, 4.

[16] Ibid., 39–40.

> or Hysterical Passions… within a few minutes brought to their senses, so as to be able to tell where their pain lay, which he has followed till he has pursued it out of the body.[17]

The affidavits collected by Greatrakes include the names of those who testify to the truth of the witness statements, but who may not necessarily have been present. Just as important is the testimony as to Greatrakes' good and pious character, and the social status of those who testify (bishops, physicians, gentry). Greatrakes makes no attempt to explain his success, beyond presenting his healing powers as a gift from God, which it is his bounden duty as a Christian to exercise, 'my duty as a Christian in imploying that Talent which God had intrusted me withal.'[18] He also says he does so as an act of charity towards those whom medicine has failed, for 'the good of his poor Fellow-creatures, whose Distempers (many of them) neither Art nor Physick probably could reach.'[19] But he refrains from any attempt to describe his technique or the responses of patients during treatment.

By contrast, in *The Miraculous Conformist*, Stubbes made a number of additional observations on the basis of the case histories which he presents. Stubbe explains that he was encouraged by Lord Conway to question Greatrakes freely. He supplies a lot of information about Greatrakes—that he was untaught in medicine or surgery, and that his healing activities were initially applied to breast cancers and the kings evil, but developed to include a whole range of diseases from falling sickness to 'old aches and pains', and that he had latterly supplemented his 'stroking' with incisions of swellings.

He notes that Greatrakes' stroking did not always work. Stubbe took pains to present particular cases in as unbiased a manner as possible. All of them were based on first-hand observation, which he followed up by interviewing those whom Greatrakes cured. Noting the importance of physical touch, he connected his 'gift of healing' to 'the temperament or composure of his body; because I see it necessary that he Touch them, or otherwise rubbe their Eyes with his Spittle'. He also noted that Greatrakes, 'Cures none wherein there is a decay in Nature', but that his treatment is efficacious where nature is 'oppressed by some Heterogeneous Ferment, either exciting an Ebullition and paine, or Coagulating and fixing the blood and spirits….'. He also noted that it was reported by reliable witnesses (Lord Conway, George Rust and Greatrakes' bedfellow Sir Amos Meredith) that Greatrakes' exuded a fragrant odour. Separately from the question of miracles, he offered an explanation of the efficacy of Greatrakes' healing powers. Stubbe, regarded the healing process as very much a *physical* process that admits of a *natural* explanation, and he offered one in terms of medical theory of fermentation, rather than in terms of physical expulsion by touch:

> It is Nature Cures the Diseases and distempers and infirmities, it is Nature makes them fly up and down the Body so as they do: they avoyd not his Hand, but his Touch and stroke so

[17] Greatrakes 1666, 61.

[18] Ibid., 41.

[19] Ibid., 3.

> Invigorateth the parts that they refect the Heterogeneous Ferment, till it be outed the Body at some of these parts he is thought to stroke it out at.[20]

He concluded that Greatrakes' capacity to cure was a gift, which he ascribed to his constitution: 'I place the gift of healing in the temperament or composure of his body'. Nevertheless he conceded that this was a gift of God

> That God had bestowed upon Mr. Greatarick a peculiar Temperament, or composed his Body of some particular Ferments, the Effluvia whereof, being introduced sometimes by a light, sometimes by a violent Friction, should restore the Temperament of the Debilitated parts, re-invigorate the Bloud, and dissipate all heterogeneous Ferments out of the Bodies of the Diseased, by the Eyes, Nose, Mouth, Hand, and Feet.[21]

Stubbe's explanation rests on a theory of the constitution of living organisms, where 'ferment' is fundamental to physiology. Thus it is possible ('imaginable') that Greatrakes' cures are produced by a 'ferment' imparted by the friction of his stroking, which revives malfunctioning parts of the body:

> Considering that our life is but a Fermentation of the Blood, nervous Liquor, and innate constitution of the parts of our Body, I conceive I have represented those hints and proofs which may render it imaginable that Mr Greataricks by his stroking may introduce an oppressed Fermentation into the Blood and Nerves, and resuscitate the oppressed Nature of the parts. I now proceed to shew that those effects which are so admired, upon the stroaking of Mr Greataricks Hand, viz: that the paines flye before his Hand, untill he drive them out at the Fingers, Mouth, Eyes, Toes, &c. are the effects of Nature in vigorated, and not immediatly of his Touching.[22]

Stubbe's recourse to fermentation theory is rendered plausible by his observation that, although physical contact was essential, it was not just the fact of Greatrakes touching his patient that effected the cure. He observed that the pain did not disappear 'immediatly of his Touching', but seemed to travel as it was chased out from different parts of the body ('Fingers, Mouth, Eyes, Toes, &c').

At this time 'fermentation' was a relatively commonplace notion in medical theory—it was an important component of Paracelsian, and Helmontian medicine, as well as a key element of mechanistic physiology. Notably, Descartes explained the circulation of the blood by means of fermentation process in the heart. One of the most important statements of medical fermentation theory in England was by one of the dedicatees of Stubbe's *Miraculous Conformist*, the Oxford physician, Thomas Willis whose *De fermentatione* was published in 1659.[23] On Willis's account,

> Fermentation is an intestine motion of Particles, or the principles of every Body, either tending to the perfection of the same Body, or because of its change into another. For the

[20] Stubbe 1666, 10–14.

[21] Ibid., 10.

[22] Ibid., 10–11.

[23] Willis's *De fermentatione* was translated into English in *The remaining medical works*, in 1681, and again, by Samuel Pordage in 1684.

> Elementary Particles being stirred up into motion, either of their own accord or Nature, or occasionally, do wonderful move themselves, and are moved.[24]

'The Inorganical Motion' of bodies that constitutes fermentation entails the refinement or 'attenuation' of larger particles by more subtle ones, the thicker particles 'by the Endeavour and Expansion of the more Subtil' are 'attenuated'. The purpose is to achieve 'perfection in the subject' or to complete alterations 'designed by Nature.'[25] Stubbe's somewhat equivocal comparison of Greatrakes' cures to the miracles of Christ scandalized many.[26] Boyle's correspondent, Daniel Coxe, interpreted Stubbe to mean that Christ's healing was naturalistic rather than miraculous.[27]

6.2 Anne Conway

One patient for whom there is no personal witness statement of Greatrakes's attempt to cure her is the person who was the principal reason for his 1665 visit to England, Lady Anne Conway. We don't know what she herself made of him as a practitioner, apart from a statement that she thought him honest, and under-valued in his own country (Ireland). It would be intriguing to know how she reacted to his failure to cure her, whether and how she explained his apparent success as a healer, or what she thought of other people's attempts at explanations. One might be tempted to think that a failure in therapeutics was bound to undermine the theories constructed in support of them. On the other hand, at a time when the state of medical knowledge was acknowledged to be imperfect, it was not so obvious that theories had to be rejected simply because of explanatory failure. Besides, the apparent success of a particular therapy (such as Greatrakes' 'stroking') in some cases was probably more likely to win endorsement for a theory which appeared to account for it.

Although her letters from the period when she encountered Valentine Greatrakes give no hint of her experience of pain, Anne Conway's correspondence contains abundant evidence of her suffering throughout her life. Given that from her late teens till her death she suffered extended bouts of unrelievable pain, which became worse as she grew older. One of the most striking things about Anne Conway's own letters is that she comes across as remarkably forbearing about giving voice to her suffering, and is almost apologetic when she mentions her 'afflictions'. Nevertheless, there is ample testimony from those around her that the pain she suffered was acute and prolonged. For example, on one occasion in 1658 she was ill for a period of over 7 weeks, during which, her husband told his brother-in-law George Rawdon, 'her sighs, and grones come so deep from her, that I am terrifyed to come neere her.'[28] As

---

[24] Willis 1681, 9.

[25] Ibid., 10.

[26] Stubbe 1666, 26–7.

[27] Boyle 2001.

[28] Lord Conway to Major Rawdon, 17th August 1658 (Hastings MSS, HA 14363).

tutor, friend and mentor of Anne Conway throughout her adult life, Henry More was well aware of her affliction, and he was put to the test to help her cope with the pain which she suffered. On the occasion just mentioned, Lord Conway reported that More stayed by to support her as best he could.[29] More was always on the look-out for a promising practitioner who might her help her—it was he who persuaded Francis Mercury van Helmont to visit her, in the hope he might be able to alleviate her pain. A less successful recommendation of a practitioner by Henry More was Samuel Hartlib's son-in-law, Frederick Clodius, who almost killed her by mercury poisoning.[30] More's standard advice to Anne Conway was to counsel patience and fortitude, reading spiritual guides like the *Theologia germanica*, and strengthening her mind with philosophy and religion. The tragedy of the loss of her only child brought Anne Conway to the limits of such consolations of philosophy, so much so that she protested her inability to take his advice. With time, an added burden was that her chronic illness became so much the norm of her condition, that she lost sympathetic support from her family and those around her, as she told Sir George Rawdon near the end of her life, 'ye greatest afflictions by long continuance grows unregarded.'[31]

The case of Anne Conway is testimony to the different conditions under which disease was experienced in the seventeenth century. In her desperate search for means to alleviate her debilitating headaches, she faced all kinds of hazards, from mercury poisoning to head surgery. Surgical options were dire: among cures for the headache described by Thomas Willis in his *De anima brutorum* (1672), there is one 'For an obstinate Head-ach[e]' which involved 'the opening of the Scull with a trepand Iron near the place of the pain'. What Willis doesn't mention is that this was performed without anaesthetic. He also mentions that no less a physician than William Harvey tried to persuade 'a[n unnamed] Lady of great quality to this, who was troubled with a terrible and inveterate Hedach' to go through with this intervention. The lady in question evidently didn't do so because he says the proposed treatment was as yet untried. (Willis himself was sceptical whether it would work). It is possible that this 'Lady of great quality' may have been Anne Conway, who was one of Willis's patients. Her case almost certainly figures elsewhere in his book, although she is not named. She did consider trepanning, and went to Paris in 1656 with a view to undergoing it. But she was dissuaded from doing so after she got there, possibly because the experts recognised that she was an unsuitable case for treatment. Closer to home, there is some suggestion that she may have submitted to cranial surgery. The evidence comes from remarks in the notebooks of John Ward, the Vicar of Stratford upon Avon, not far from Ragley Hall. In notes taken between 1663 and 1665, Ward records that 'The Lady Conway hath great pains in her head her sutures open'. He adds, a few lines later, 'that shee came into this miserie by a Brother of hers who was a Traveller and had some skill in physick and chymistrie and tried

[29] Hastings MSS, HA 14363.

[30] Nicolson and Hutton 1992, 94ff.

[31] Ibid., 448.

experiments uppon her.'[32] However, there is no corroboration of an attempt by her medically-qualified brother, Sir John Finch to operate on her.[33] Ward's remarks suggest that there had been some procedure performed on her head. By contrast, Valentine Greatrakes' non-invasive techniques must have seemed a very balm.

We have no way of knowing whether her experience at the hands of Greatrakes had any bearing on how her philosophy evolved. It has become almost a truism of Conway studies that her denial that soul and body are distinct substances monism is connected to her personal experience of pain—after all to understand states of mind as the same as states of body is a plausible way of explaining why the mind should suffer when the body is in pain. While personal experience must be a factor in anyone's perspective on life, it is an oversimplification to say that she built a theory on her personal misfortune of suffering lifelong chronic pain. That is tantamount to saying that she drew the short straw in life and tried to explain it away by writing a treatise of metaphysics. There is certainly more to her understanding of pain than simply a response to Cartesian dualism.

That said, there is no denying that for Conway, pain and suffering are inexplicable in dualistic terms: one of Anne Conway's 'proofs' that spirit and body are not essentially distinct from one another turns on the experience of pain.

> …why does the spirit or soul suffer so with bodily pain? For if when united to the body it has no corporeality or bodily nature, why is it wounded or grieved when the body is wounded, whose nature is so different….

Difficult questions like this about the interaction of soul and body lead her to repudiate soul-body dualism, positing instead substance monism, which meant that states of soul and body are not distinct from one another.

> But if one admits that the soul is of one nature and substance with the body, although it surpasses the body by many degrees of life and spirituality, just as it does in swiftness and penetrability and various other perfections, when all the above mentioned difficulties vanish, and one may easily understand how the soul and body are united and how the soul moves the body and suffers with it and through it.[34]

Thus, in Conway's view, pain can only make sense in non-dualist terms. Descartes had argued in his *Meditations* that sensation is a key indicator that bodies exist; and that pain as an extreme form of bodily discomfort, is the most direct evidence that we have bodies as well as minds. Anne Conway would have agreed with Descartes that pain makes us aware of our bodies, but not with his conclusion that this confirms that the body is different from the soul/mind. She argued to the contrary, that we feel pain because bodies and souls are *not* distinct entities.

---

[32] Ward 1839, 549 and 551–2.

[33] Hutton 2004, chapter 5.

[34] Conway 1996, 58. Like her teacher, Henry More, she clearly did not think that Descartes could satisfactorily explain soul-body interaction. But Conway's critique differed from More's fundamentally (Hutton 1995, 2004).

> … And since the body itself is sentient life or perceiving substance, it is no less easily understood how one body can or would bring pain or pleasure to another body, because things of one or of a similar nature can easily affect each other….[35]

In this way Conway inverts Descartes' argument for soul-body dualism.

But there is more to her discussion of pain than a critique of Cartesian (and Morean) dualism, for pain plays a central role in the moral and ontological aspect of her metaphysical system. Another feature of Conway's account of pain is that she links it to the condition of corporeality itself. In her metaphysics, corporeality is the state of imperfect beings which have fallen away from their original pristine goodness and spiritual perfection. This involves a change of state or mode, not a change of substance. The further beings are from their original perfection, the more corporeal their constitutive substance ('gross' or 'crass' in the English translation). In this process, pain is associated with the physical changes involved.

> If one asks whether these spirits become more corporeal than they previously were in their original state before they fell through their own wrongdoing, I answer, yes…[36]

However, pain is not merely the by-product of the physical condition of created beings. Pain is purposeful in the grand scheme of things, and purposeful in two respects. First it is a punishment for sin—to that extent, she retains a traditional view. But, rather less traditionally, she holds that the purpose of punishment is not merely retributive, but that it is beneficial: 'all these punishments tend toward the good of the creatures'. Indeed she goes further, to claim that pain is *essential* for achieving the restoration of the spiritual health and being of creatures.

> all pain and torment stimulates the life or spirit existing in everything which suffers. As we see from constant experience and as reason teaches us, this must necessarily happen because through pain and suffering whatever grossness or crassness is contracted by the spirit or body is diminished; and so the spirit imprisoned in such grossness or crassness is set free and becomes more spiritual and, consequently, more active and effective through pain.[37]

Furthermore, the more the suffering the quicker the recuperation: 'the greater its suffering, the sooner its return and restoration.'[38] Thus, through the role of pain and suffering in this process of restoration, Conway takes the moralisation of pain to an extreme. What we have here is the basis of what D.P. Walker called Conway's 'terrible theory' of the 'extreme moral value of suffering', mitigated only by her adoption of an extreme theodicy which entails repudiation of the doctrine of hell as a state of post-mortem punishment, and subscription to the idea of universal salvation (*apocotastasis*).[39] Conway's position is explicable in terms of an attempt to reconcile her terrible experience of unrelievable pain with the conception of God's providence.

---

[35] Conway 1996, 58–9.

[36] Ibid., 43.

[37] Ibid.

[38] Ibid.

[39] Walker 1964, 139–40; Hutton 2012.

Moralisation aside, there are two particular features of Conway's account which I want to highlight: first of all her adoption of a medical metaphor for the process of regeneration through pain, and the role of Christ. According to Conway, the process of restoration is a healing process, in which 'all Punishments are Medicinal.'[40] And the restoration of all things to their original purity is an operation which is made possible through Christ as 'middle nature' acting as 'a most powerful and efficacious *balm*' (my italics). This process, which restores things from their fallen state to original perfection, recapitulates the pattern of redemption-through-suffering exemplified by the biblical Christ. Anne Conway draws an analogy between the pains suffered in this process and the sufferings of the biblical Christ by which he secured the redemption of mankind. The work of the incarnation, she writes, entails that Christ 'subjects himself to suffering and death… so that he may raise the souls of men above time and corruption up to him himself, in whom they receive blessing and in whom they grow by degrees in goodness, virtue and holiness forever.'[41] This work of redemption, too, is described as a healing process. 'Christ', she notes, 'suffered great torment and death itself' in order to 'heal, preserve, and restore creatures from corruption and death'. Conway presses the medical analogy by comparing the healing process of the work of redemption to the operation of a 'ferment':

> In assuming flesh and blood, he sanctified nature so that he could sanctify everything, just as it is the property of a ferment to ferment the whole mass.'[42]

While Conway clearly invokes Christian doctrine here, we should bear in mind that it is not the historical Christ that she has in mind, so much as the metaphysical 'Christ' of her philosophical system, which she also identifies as a 'middle nature', and 'Adam Kadmon'. The medical analogy which Anne Conway employs is not accidental. For the restoration in question is not just a matter of moral regeneration, but it involves the purifying of the substance of which all beings are constituted. Conway describes the healing/restoration process as one in which the spiritual core of bodies is liberated by reducing their 'grossness and crassness':

> whatever grossness or crassness is contracted by the spirit or body is diminished; and so the spirit imprisoned in such grossness or crassness is set free and becomes more spiritual …[43]

This recalls Willis's account of fermentation, which, as we noted earlier, involved refinement of thicker particles by more subtle ones. The thicker particles are broken down ('attenuated') 'by the Endeavour and Expansion of the more Subtil', in order to achieve 'perfection in the subject' or to 'complete alterations designed by Nature.'[44] Thus Anne Conway's account of the workings of pain perhaps owes as

40 Conway 1996, 38.

41 Ibid., 27.

42 Ibid. The reference to ferment is lost in the 1690 English translation where the Latin 'tanquam pars fermenti ad fermendam totam massma' is translated as 'so was as that little Leaven that changed the whole Lump' (Conway 1998, 18 and 129).

43 Conway, 1996, 43.

44 Willis 1681, 10.

much to medical as to moral theory—so much so that much of her treatise could be described as a moral pathology.

6.3 Stubbe and the Conway Circle

As already noted, fermentation is a component of the medical theory of one of the witnesses of Greatrakes's strokings, the much-maligned, Henry Stubbe. Conway's comparison of Christ's healing, and of the efficacy of pain in breaking down hard substance bears to the process of fermentation has some striking analogies with the fermentation theory advanced by Stubbe. In pointing out this parallel, my claim is not that Stubbe is the source of Conway's medicalised account of redemption. But her account of the restorative efficacy of pain, and the healing powers of Christ are certainly compatible with it, and it is striking that this positive outcome (relief from pain) is achieved through fermentation, in a manner not dissimilar to how Stubbe explains the efficacy of Greatrakes' touch.

The reputation which Stubbe has acquired among historians militates against any suggestion that Conway may have been receptive to Stubbe.[45] However, while there is no doubting that Stubbe would become a gadfly to the Royal Society arguably, focusing on these disputes has given a distorted impression of Stubbe and what he represented. Most discussions of Stubbe focus on Boyle and the Royal Society. This is understandable in view of Boyle's standing in histories of Royal Society science and his known displeasure at Stubbe's account of the Greatrakes' affair, which was addressed to Boyle. Stubbe's quarrel with Glanvill, in particular, has coloured assessments of him. In the course of that quarrel he turned on Henry More, who had chosen to support Glanvill and did so publicly.[46] Discussions of Stubbe and the Royal Society tend to take Stubbe's detractors at their word. As a result, Stubbe's observations on Greatrakes tend to be read in the light of his subsequent anti-Royal Society polemics. There is a dominant view that Stubbe deliberately courted controversy and that he intended offence to Boyle. He is even seen as advancing a radical political agenda (Elmer, Jacob) and as using Greatrakes in a deliberate move to undermine religion. With the exception of Harold J. Cook, the fact that Stubbe had medical credibility among his contemporaries has been largely forgotten, and the value of *The Miraculous Conformist* as an exercise in medical diagnosis has been overlooked.

---

[45] For the historian's perspective, see Kaplan 1982; Steneck 1982; Jacob 1983; Cook 1989; Elmer 2013. These works of Elmer, Jacob and Cook are referred to again later in the paragraph above.

[46] The reason for Stubbe turning on More was not indiscriminate polemical petulance, but out of a sense of betrayal. He essentially accused More of bad faith in siding with Glanvill and the Royal Society, having agreed with Stubbe (privately) that he shared his concerns about the atheistical tendencies of the mechanical philosophy. Stubbe's polemics against Glanvill in fact contain the most devastating critique of More by any of his contemporaries.

Stubbe certainly stood for an alternative methodology to the 'Baconianism' of the Royal Society. Instead of appealing to Bacon, he invoked a medical tradition, deriving from Hippocrates and Galen, which also valued Aristotle, but not for the truth of his doctrines, rather for his attempt to found his philosophy on sensory observation. A very different view of Stubbe from the subversive reputation he has acquired can be had from others who knew him, especially in a medical capacity. One such was James Cook, translator of John Hall's *Select Observations on English Bodies of Eminent Persons in Desperate Diseases* (1679). In his Preface to this book (dated 25/3/1679), Cook, remembers Stubbe both as a person and as a practitioner in very positive terms, recalling Stubbe's 'Civilities and Openness', and stating that, 'He was not a Person greedy of Gain. Not many months before he died, he said to me, "We must study all ways possible to find out and appoint Medicines of cheap rate, and effectual; for Money is scarce, and Country-People poor."'[47] When James wrote this, Stubbe was long dead, so we may be sure that James had no interest at stake in expressing his approbation of him.

I would argue that, contrary to the view that Stubbe intentionally courted trouble at the time of the Greatrakes affair, the views which he expresses in *The Miraculous Conformist* deserve to be taken at face value as the observations of a practising physician. It is more than likely that he thought he would get a good reception from Boyle as from Willis—though he may well not have expected them to agree with him. Boyle was certainly displeased that Stubbe had addressed *The Miraculous Conformist* to him without permission, and he criticised Stubbe on the question of miracles. Nevertheless, he accorded Stubbe the courtesy of a long reply which addressed his observations point by point. He did not object to naturalistic explanations for Greatrakes' cures: it was, he said 'not unlawful to endeavour to give a physical account of his cures and to enquire whether his touch be any more than a more noble specific, that reaches not to all diseases, or to most, but to more than the generality of specificks.'[48] But Boyle declined to commit himself to any particular explanation, professing himself, 'very shy of admitting any thing as a principle in matters purely physical, that I cannot well understand, and do not find well proved.'[49]

There is also compelling evidence that members of Anne Conway's circle were receptive to Stubbe's views on medicine and natural philosophy. First, Stubbe evidently retained the Conways' confidence in him as a physician after Valentine Greatrakes' unsuccessful attempt to cure Lady Conway. He advised on treatment of Lord Conway's nephew in 1671, and when Stubbe was drowned in a riding accident, Lord Conway took steps to secure his estate on behalf of his mother.[50] Stubbe's

[47] Hall 1679, Preface.

[48] Boyle 1772, 1: lxxi. It is not clear whether the letter was sent (Hunter 2009). For Boyle's 'Accounts of cures performed by Valentine Greatrakes during his visit to England in 1666' in his Work Diary see WD-26 electronic edition http://www.livesandletters.ac.uk/wd/view/view.html

[49] Boyle 1772, 1: lxxi.

[50] Hastings MSS HA 14550, 14,551. Viscount Conway expressed an interest in acquiring Stubbe's books, noting that he had 'a very great Library and very choice' (Huntington Library Hastings MS HA 14554).

contact with the Conways put him in the orbit of a circle of thinkers who, like himself, were critical of the mechanical philosophy of Descartes. These were Lady Conway herself, Henry More and also her half-brother, Sir John Finch. Stubbe may have met Finch when he visited Ragley Hall in 1671.[51] This was a circle where the medical researches of Thomas Willis, dedicatee of *The Miraculous Conformist*, were favourably received. Not only had Willis attended Lady Conway as physician, but her hall-brother, Sir John Finch was an anatomist with high regard for Willis. Finch had made a special study of the anatomy of the brain, and been instrumental in introducing Willis's works to the anatomists in Italy.[52] Furthermore, in Finch, Stubbe would have found someone who shared the view that there was an alternative experimental approach to that of the Royal Society—one which could be traced back to the great medical minds of antiquity, Hippocrates and Galen. This, too, was an approach which relied on observation and the evidence of the senses. And, as Stubbe pointed out Aristotle, too, had adopted such an approach in his biological writings. Finch, like Stubbe, was well aware of alternative experimental approaches to natural philosophy those pioneered by the Royal Society.[53] In Finch's case it was the approach of the Florentine Accademia del Cimento with which he was connected.[54] In his *Legends No Histories* Stubbe makes reference to an academy for experiments in Florence, under the patronage of the Duke of Florence, obviously a reference to the Accademia del Cimento. It so happens that Stubbe and Finch also had other things in common, most notably a positive view of Hobbes, and a relatively enlightened view of Islam.[55] Against this background, it is not beyond the bounds of possibility that Anne Conway too might have been receptive to Stubbes's explanation of Greatrakes' healing powers.

Whether this receptivity extended to her adopting Stubbe's medical theory is another question. Stubbe was not the only possible source for fermentation theory. Willis, whom she consulted, was one. Jan Baptiste van Helmont, whose son Francis Mercury treated her, was another. And we do know that her *Principia philosophiae* reflects the impact of her encounter with Francis Mercury van Helmont. A possible attraction of Francis Mercury van Helmont's medical views, was his psycho-somatic explanation of illness, according to which ailments are to be treated by altering the attitudes and feelings of the patient: if, as Van Helmont held, diseases are the result of disordered passions (literally *dis-tempers*), they should be treated by bearing pain patiently and learning to love the disorder.[56]

---

[51] Hastings MS HA 14506.

[52] Hutton 2004, chapter 5.

[53] Finch MS, DG7 Lit, 194–5.

[54] The experimental approach of the Accademia del Cimento, though not John Finch, is discussed in Boschiero 2007.

[55] Hutton 2004, chapter 5.

[56] Van Helmont 1694, Sherrer 1958; also Hutton 1996, 2004. This was one feature of his clinical method which, according to Van Helmont, did not meet with the approval of most doctors of his time. He says his method 'would oblige them to abandon their accustomed method, and to new-model the whole system of their practise' (Helmont 1694, sig. $A4^5$).

We don't know when Anne Conway arrived at her extreme view of the moral value of pain. It is unlikely that the ministrations of Valentine Greatrakes were a catalyst. But there is no question that her philosophy reflects her acquaintance with medical theory. And it is notable that fermentation theory informs her account of the refinement of corporeal substance which is at the heart of her theory of restoration of all things. Whether or not Anne Conway owed anything to Stubbe beyond the professional services of a family physician, the Conway connection, highlighted by the Greatrakes affair, opens up the possibility of a reassessment of Stubbe. Taken in conjunction, the witness statements of Greatrakes' patients and Anne Conway's correspondence provide a rich inter-penetration of the social, cultural and intellectual contexts, in which both practitioners and patients tried to make sense of their world.

Bibliography

Manuscripts

San Marino, Huntington Library. Hastings MSS.
Leicester, Leicester and Rutland County Record Office. Finch Mss. MS DG7 Lit

Primary and Secondary Sources

Anon. 1666. *Great cures and strange miracles performed by Mr Valentine Gretrux*. London.
Anstey, Peter. 2010. The body as object and instrument of knowledge. In *Embodied empiricism in early modern science*, ed. Charles T. Wolfe and Ofer Gal, 93–117. Dordrecht: Springer.
Aucante, Vincent. 2006. *La philosophie médicale de Descartes*. Paris: Presses universitaires de France.
Bacon, Francis. 1640. *Of the advancement and proficience of learning*. London: Leon Lichfield.
Boyle, Robert. 1772. *The works of the honourable Robert Boyle*, new edition, ed. Thomas Birch. London: J. and F. Rivington.
———. 2001. *The correspondence of Robert Boyle*, ed. Michael Hunter, Antonio Clericuzio, and Lawrence M. Principe, 6 vols. London: Pickering & Chatto.
———. 2005. *Robert Boyle's 'heads' and 'inquiries'*, ed. Michael Hunter. Boyle Project Occasional Papers 1. London: Birkbeck, University of London.
———. 2007. *Work diary*, ed. Michael Hunter et al. London: CELL. http://www.livesandletters.ac.uk/wd/view/view.html.
Boschiero, Luciano. 2007. *Experiment and natural philosophy in seventeenth-century Tuscany: The history of the Accademia del Cimento*. Dordrecht: Springer.
Brogan, Stephen. 2015. *The royal touch in early modern England*, Politics, medicine and sin. Woodbridge: Boydell Press.
Broad, Jacqueline. 2003. *Women philosophers of the seventeenth century*. Cambridge: Cambridge University Press.
Conway, Anne. 1690. *Principia philosophiae antiquissimae et recentissimae*. Amsterdam.
———. 1996 *The principles of the most ancient and modern philosophy*. Translated by Allison P. Coudert and Taylor Corse. Cambridge: Cambridge University Press.

———. 1998. In *The principles of the most ancient and modern philosophy*, ed. Peter Loptson, 2nd ed. Delmar: Scholars' Facsimiles and Reprints.

Cook, Harold J. 1989. Physicians and the new philosophy: Henry Stubbe and the Virtuosi-Physicians. In *The medical revolution of the seventeenth-century*, ed. Roger French and Andrew Cunningham, 246–217. Cambridge: Cambridge University Press.

Crocker, Robert. 2005. 'A sanative contagion'. Henry More and faith healing. In *Practice of reform in health, medicine and science, 1500-2000*, ed. S. Mandelbrote and M. Pelling, 107–123. Aldershot: Ashgate.

Descartes, René. 1984–1991. *The philosophical writings of descartes*. Translated by John Cottingham, Dugald Murdoch, Robert Stoothoff, and Anthony Kenny, 3 vols. Cambridge: Cambridge University Press.

Dewhurst, Kenneth. 1963. *John Locke (1632-1704), physician and philosopher: A medical biography*. London: Wellcome Library.

DiMeo, Michelle. 2015. 'Such a sister became such a brother': Lady Ranelagh's influence on Robert Boyle. *Intellectual History Review* 25: 21–36.

Elmer, Peter. 2013. *The miraculous conformist: Valentine Greatrakes, the body politic, and the politics of healing in restoration Britain*. Oxford: Oxford University Press.

———. 2004. Greatrakes, Valentine. In *Oxford dictionary of national biography*. Oxford University Press. http://www.oxforddnb.com/view/article/11367. Accessed 1 Nov 2015.

Frank, Robert G. 1974. The John Ward diaries: Mirror of seventeenth-century medicine. *Journal of Medical History* 29: 147–179.

Hall, John. 1679. *Select observations on English bodies of eminent persons in desperate diseases*. London: Benjamin Shirley.

Helmont, Francis Mercury. 1694. *The spirit of diseases: Or diseases from the spirit laid open in some observations concerning man, and his diseases, wherein is shewed how much the mind influenceth the body in causing and curing of diseases*. London: Sarah Howkins.

Hunter, Michael. 2009. *Boyle: Between god and science*. New Haven: Yale University Press.

Hutton, Sarah. 1995. Anne Conway, critique de Henry More: substance et matière. *Archives de Philosophie* 58: 371–384.

———. 1996. Of physic and philosophy. Anne Conway, Francis Mercury van Helmont and seventeenth-century medicine. In *Religio Medici. Medicine and religion in seventeenth-century England*, ed. Andrew Cunningham and O. Grell, 218–246. Aldershot: Scolar Press.

———. 2004. *Anne Conway: A woman philosopher*. Cambridge: Cambridge University Press.

———. 2012. Anne Conway und die Güte Gottes. In *Anne Conways 'Principia philosophiae': Materialismuskritik und Alleinsheits-Speculation im neuzeitlichen England*, ed. W. Weichert and C. Hengstermann, 67–83. Berlin: LIT Verlag.

Jacob, James R. 1983. *Henry Stubbe, radical protestantism and the early enlightenment*. Cambridge: Cambridge University Press.

Kaplan, Barbara. *1982*. Greatrakes the Stroker: The interpretations of his contemporaries. *Isis* 73: 178–185.

Manning, Gideon. 2007. Out on the limb: The place of medicine in descartes' philosophy. *Early Science and Medicine* 12: 214–222.

———. 2014. Descartes and the Bologna Affair. *British Journal for History of Science* 47: 1–13.

Milton, J.R. 2001. Locke, medicine and the mechanical philosophy. *British Journal for the History of Philosophy* 9: 221–243.

More, Henry. 1690. *An account of virtue*. This is a translation of his *Enchiridion ethicum*. London: Benj. Tooke.

Nicolson, Marjorie Hope, ed., revised by Sarah Hutton. 1992. *The Conway letters: The correspondence of Anne, Viscountess Conway, Henry More, and their friends: 1642-1684*. Oxford: Clarendon Press.

Shaw, Jane. 2006. *Miracles in enlightenment England*. New Haven/London: Yale University Press.

Sherrer, G. 1958. Philalgia in Warwickshire: F.M. van Helmont's anatomy of pain applied to Anne Conway. *Studies in the Renaissance* 5: 196–206.

Steneck, Nicholas. 1982. Greatrakes the Stroker: The interpretations of historians. *Isis* 73: 161–177.

Stubbe, Philip. 1666. *The miraculous conformist: Or an account of several marvailous cures performed by the stroaking of the hands of Mr Valentine Greatarick, with a physical discourse thereupon, in a letter to the honourable Robert Boyle Esq*. Oxford: H. Hall.

Villani, Stefano. 2005. Between anatomy and politics: John Finch and Italy, 1649-1671. In *Practice of reform in health, medicine and science, 1500-2000*, ed. S. Mandelbrote and M. Pelling, 151–166. Ashgate: Aldershot.

Walker, D.P. 1964. *The decline of hell. Seventeenth-century discussions of eternal torment*. London: Duckworth.

Ward, John. 1839. In *Diary of the Reverend John Ward, A.M., Vicar of Stratford-upon-Avon, extending from 1648-1679*, ed. C. Severn. London: Henry Colburn.

———. 2000. *The life of the Pious and Learned Henry More*, New edition, ed. S. Hutton et al. Dordrecht: Kluwer.

Willis, Thomas. 1681. *The remaining medical works*. London: T. Dring, C. Harper, J. Leigh, and S. Martyn.

Chapter 7
Animal Bodies and Human Minds: The Anatomy of the Brain and the Analogy of Nature

Anita Guerrini

Abstract How did the widespread and lasting assumption of an analogy between animal and human bodies affect ideas about the mind-body relationship? In particular, how did seventeenth-century men of science justify their employment of animal brains to talk about human brains and minds? Galenic ideas of the structure of the brain, based on animals, gave way to a multiplicity of ideas about the relationship between the brain and the mind (or soul). The continued reliance on animal models would seem to conflict with a new emphasis on specifically human structures and behaviors. But at the same time anatomists and philosophers also looked anew at animal cognition. The work of Marin Cureau de la Chambre, Niels Stensen, and Thomas Willis reveals some of these new ideas and the tensions that resulted.

Keywords Brain anatomy · Soul · Animal models · Analogy · Animal cognition · Marin Cureau de la Chambre · Niels Stensen · Thomas Willis

7.1 Introduction

The use of animal bodies as stand-ins for humans in anatomical study has a long history.[1] How did the widespread and lasting assumption of an analogy between animal and human bodies affect ideas about the mind-body relationship? This analogy was, indeed, so ingrained in early modern science that it became akin to a paradigm, a guiding principle of research. In particular, how did seventeenth-century

[1] von Staden 2013.

A. Guerrini (✉)
Oregon State University, Corvallis, OR, USA
e-mail: Anita.Guerrini@oregonstate.edu

G. Manning (ed.), *Testimonies: States of Mind and States of the Body in the Early Modern Period*, Archimedes 57,
https://doi.org/10.1007/978-3-030-39375-5_7

men of science justify their employment of animal brains to talk about human brains and minds? I will begin with a short overview of this idea of analogy and how it was applied before moving on to a few specific applications of this concept in the seventeenth century.

Most historians have tended to see the mind-body relationship with regard to animals through a Cartesian lens. The usual argument has been that the famous comments on animal cognition in *Discourse on the Method* described Descartes's belief that animals did not possess "mind" as such, in terms of rational thought; he wrote, "it is nature which acts in them according to the disposition of their organs," comparing animals to a clock.[2] Therefore, the argument continues, most seventeenth-century investigators operated on the assumption that animals could not think or feel. But in fact many authors did not share Descartes's assumptions about the mind-body relationship and about animal cognition, and it's not clear that even Descartes was as "Cartesian" as we have been led to believe. This essay will look at some of these non-Cartesian (and even anti-Cartesian) ideas, particularly the voluminous writings of the French physician Marin Cureau de la Chambre between the 1640s and the 1660s and the less voluminous but still substantial works of the English physician Thomas Willis between the 1650s and the 1670s. Willis was one of a number of English physicians who carried on the research program of comparative anatomy, based on extensive dissection of a wide variety of living and dead animals, begun by William Harvey.[3] Contemporary with Willis are the brain studies of Niels Stensen (Nicolaus Steno). Their views on mind, soul, and animal cognition differed significantly from each other and from Descartes, during a period when, as Fernando Vidal has described it, soul was coming to be identified with mind, and mind was coming to be localized in the brain, so that selfhood became what Vidal calls "brainhood."[4] I would argue, however, that this process was still ongoing during the period in question. Cureau de la Chambre's work on animal feeling and cognition was based on philosophical principles and behavioral observation, while Willis employed comparative anatomy and mechanistic chemistry to talk about what he called the rational and animal souls. Stensen avowed himself to be a mechanist, but his anatomical studies of the brain criticized Cartesian ideas of the structure of the brain and its relation to animal cognition. He did not, however, offer an alternative theory. None of these men questioned the validity of the analogy between human and animal bodies, but they reached different conclusions about its significance.

[2] Descartes 1985, I:141.

[3] Frank 1980.

[4] Vidal 2009.

7.2 Animals and the Analogy of Nature

Aristotle's gradualist theory of classification assumed a chain of being with incremental spaces between species and between genera. This in turn was based on the assumption that there is an underlying shared essential nature (*physis,* the principle of growth) in all living things.[5] Humans, to Aristotle, were another animal and shared structures with other animals, particularly mammals. His *Historia animalium* classifies humans and viviparous quadrupeds as a single group, and Book II of that work details the numerous similarities, differences, and analogies among them.[6] He defined analogy in the first book of *Historia animalium* in the context of differentiating animals. "Genus" referred to larger categories such as birds or fish; there were in turn many species of each genus. Among species, "the differences exhibited in animals by most of the parts lie in the contrasting oppositions of their secondary characteristics" such as color or shape. These differences can be viewed as "the more and less" or "excess and defect." Parts that cannot be defined in this way (i.e. as either identical or "more and less") are similar by way of analogy. Aristotle gives as examples nail and hoof, hand and claw, fish scale and bird feather.[7] Aristotle referred many times in *De partibus animalium* and *Historia animalium* to "analogy" and "analogous" structures. Such parts (and they are always parts, not whole animals) can be identified by external observation. But as these examples show, what Aristotle meant by "analogy" was not the same as the modern definition of a correlation or a resemblance, because it was specifically based on use rather than on a general morphological comparison; it is closer to what we now call "homology," particularly "general homology."[8] D.M. Balme referred to Aristotelian analogy as "equivalent parts as judged by function."[9]

This emphasis on what I might call "functional analogy" led Aristotle, and many after him, to find no logical problem in viewing humans as essentially animals and yet finding a fundamental distinction between human and animal on the basis of mind. As Peter Harrison has noted, "the chain of being provided the ontological basis for knowledge based on analogy," and the widely-employed concept of the microcosm and macrocosm can be viewed as an analogy.[10] But a functional analogy between animal and human hearts or even brains did not preclude the possibility of other functions apart from the analogous ones; although the things being compared may be identical in one or more respects, this is no guarantee they are identical in all respects. Thus in *De generatione animalium,* Aristotle claimed a functional

---

[5] Atran 1987.

[6] Aristotle 1965; see Lloyd 1961.

[7] Aristotle 1965, 486a14-b22; cf. Wilson 1997, 335–36.

[8] Wilson 1997 gives a good survey of the interpretive issues surrounding Aristotle's concept of analogy, which I will not repeat here.

[9] Balme 1972, 148.

[10] Harrison 1998, 465.

analogy between semen and menstrual blood, while acknowledging that they are quite different substances.[11]

The idea that human bodies largely functioned like animal bodies, and that specific parts were functional analogues to those of certain animals, became widely accepted in Western science. At the same time, in G.E.R. Lloyd's words, Aristotle's epistemology and theory of causes "not only justifies, but obliges him to undertake, the study of the whole of the animal kingdom."[12] This study included extensive dissection of dead and possibly also of living animals, and this too became a model for Western inquiry, although it was not systematically followed, with a few exceptions, until the sixteenth century. But while Aristotle believed that the study of animals was in itself of value without direct reference or comparison to humans, most subsequent researchers viewed animals primarily as surrogates for the human body.

The principle in Western science that certain animals could legitimately act as stand-ins to examine the inner parts of human bodies owed as much to Galen as to Aristotle. Like Aristotle, Galen, 500 years later, lived in a culture that frowned upon human dissection. Therefore he relied entirely on animal anatomy to describe the human body, and his anatomical works display little interest in animals as animals. He employed animals that were morphologically most similar to humans, particularly Barbary apes and other primates, and secondarily pigs, whose internal organs closely resemble the human.[13] Galen's use of animals as the basis of his detailed account of human anatomy and physiology reinforced and strengthened the identity between human and animal bodies, making a virtue of a necessity. Dissection of any kind largely ceased in the West not long after Galen's death in the third century C.E.; when it recommenced almost a millennium later, the pig was the preferred anatomical subject.[14]

7.3 Dissecting the Brain Before the Seventeenth Century

The Hippocratic treatise "On the Sacred Disease" described opening the head of a goat dead from epilepsy to demonstrate its natural origins: the brain appeared wet and phlegmy.[15] The Alexandrian physician Herophilus apparently dissected human brains as well as animal brains, primarily of the ox, and made several significant discoveries.[16] Galen dissected many brains in living and dead apes and other animals to demonstrate the localization of certain functions. The distinction between mind and soul, and between perception and understanding, had vexed philosophers

[11] Balme 1972, 147–48.

[12] Lloyd 1975, 142.

[13] Galen 1956.

[14] *Anatomia Cophonis* 1927.

[15] Hippocrates 1978.

[16] von Staden 1989, 157–58.

from the very beginnings of Greek thought.[17] Unlike Aristotle, who had located the soul in the heart (because he believed it to be hotter and therefore more perfect than the brain), Galen associated imagination, intellect and memory – functions we think of as belonging to mind – within the brain, although he did not localize these functions. Borrowing from Plato and from other philosophers, Galen enumerated three forms of soul – rational, spirited, and appetitive, which corresponded to three types of *pneuma* or spirit, each of which originated in a different organ. The animal spirit corresponded to the rational soul and originated in the brain; the vital spirit to the spirited soul, originating in the heart, and the natural or physical spirit originated in the liver and corresponded to the appetitive soul.[18] Galen did not discuss these matters in the context of his dissection of the brain, but in his works on therapeutics.

In his instructional manual *On Anatomical Procedures,* Galen recommended the brain of an ox as suitable for dissection, noting these were easily obtained from butchers with the skull already removed. In *On the use of the parts,* Galen described the network of vessels at the base of the skull known as the *rete mirabile,* which filtered the animal spirits out of the vital spirits produced in the heart; animal spirits therefore did not exactly originate in the brain but flowed into it.[19] As first Berengario da Carpi and then Vesalius discovered, cattle and sheep possess the *rete mirabile* but humans do not.[20] Galen assumed that ox and human brains were similar enough that one could stand in for the other. It's not clear if therefore he believed that the ox possessed animal spirits and a rational soul. As Sebastian Pranghofer has recently documented, the *rete mirabile* did not quietly go away after Vesalius, as we shall see further below.

A few others had published accounts of their dissections of the human brain before Vesalius, including Berengario da Carpi and Johannes Eichmann, known as Dryander, whose *Anatomia capitis humani* appeared in 1536. Charles Estienne's account of the brain in his *Anatomia corporis humani* (1545) also predated Vesalius although it was published after *De fabrica.* Most historians agree, however, that Vesalius's account far surpassed these. Vesalius began by asserting that the brain is made for what he called the "Reigning Soul" (*princeps anima,* a phrase he borrowed from Galen), "by which we imagine, meditate, and remember."[21] He concurred with Galen's three kinds of spirit, but he believed that vital spirit from the heart mixed with inhaled air in the brain and "by power of the peculiar substance of the brain" became animal spirit which was then distributed through the body by the nerves, dispensing with the refining function of the *rete mirabile*.

Vesalius wrote, "I can in some degree follow the brain's functions in dissections of living animals, with sufficient probability and truth."[22] He readily admitted the

[17] See Dolan 2007, Vidal 2009.

[18] Hankinson 1991.

[19] Galen 1956, 226–27, 1968, 430–34; Finger 1994, 16–17.

[20] Pranghofer 2009.

[21] Vesalius 1952, 3.

[22] Vesalius 1952, 4.

close similarities among the brains of the animals he dissected (sheep, goat, ox, cat, ape, dog) and between them and the human brain. The main difference, he said, was size: the human brain was by far the largest, followed by the ape and the dog. This order, he said, followed "the order we have learned of the power of reason in animals."[23] Although his dissection of the brain purported to describe a human brain, he frequently mentioned animals. The cerebrum, for example, was equally convoluted (as Galen had noted) in both humans and other animals; Vesalius surmised that the folds may be deeper in humans, "that the substance of the brain may be the richer," but he did not pursue this idea and agreed with Galen that neither convolutions nor the quantity of animal spirits bore any relationship to the level of intelligence (Vesalius 1952, 24–25).

Vesalius pointed out that the parts of the brain that made animal spirits existed in all animals (or at least the ones most commonly dissected), and that there was no special part that existed in human brains alone but not in animals that could be responsible for the "reigning soul." And indeed Vesalius never quite came out and said that animals did not have reason. Although he located the creation of animal spirit in the "ventricles" of the brain, he added, "this explains nothing about the faculties of the Reigning Soul." "All our contemporaries," he went on, "deny to apes, dogs, horses, sheep, cattle and other animals, the main power of the Reigning Soul … and attribute to man alone the faculty of reasoning." But, Vesalius asserted, there is no anatomical reason for this, since human and animal brains are largely similar. He extricated himself from this impasse with the conclusion that God gave humans alone a "Rational Soul" that is entirely spiritual and therefore cannot be found by means of dissection.[24]

Although human uniqueness remained a religious and metaphysical imperative, anatomists in the seventeenth century continued to employ animals to talk about human function, even in an area as increasingly connected to specifically human qualities as the brain. At the same time that anatomists emphasized the unique qualities of humans, they employed animals to illustrate them. The great French anatomist Jean Riolan declared that "The subject of anatomy is the body of man, the composition and conformation of which is shown by a very precise dissection."[25] Despite this emphasis on the human body, the title page of the French version of his major anatomical work *Anthropographia* described it as encompassing the anatomy of "men, women, children, and living animals." In his text, Riolan repeated with approval Galen's advice to dissect both the dead and the living, explaining the usefulness of animals in this regard, and he also dissected dead animals. His contemporary André du Laurens constantly referred to "man and other animals," even in his discussion of the brain, which he called the "true seat of the soul."[26] The assumption that form dictated function – that every structure in the body must have a functional

[23] Vesalius 1952, 6–7.

[24] Vesalius 1952, 39–40.

[25] Riolan 1628–29, 91.

[26] Du Laurens 1600, 531.

purpose – was a religious as well as a philosophical conviction, proof of the design of nature.[27]

Descartes's strict separation of mind from body and human from animal may have clarified a distinction that had been observed mostly in the breach, but he made no attempt to prove this anatomically; quite the opposite. In 1632, he commented in a letter to Marin Mersenne that "I am now dissecting the heads of various animals, so that I can explain what imagination, memory, etc. consist in." In the very next sentence he noted that he had just read Harvey's *De motu cordis.*[28] Harvey's methods clearly served as inspiration for such work. In *Discourse on the Method* he quite happily used the heart and lungs of a cow to illustrate the circulation of the blood, while asserting emphatically that animals were only body and lacked the rational soul that humans possessed. He definitively identified that soul with mind. His proof of this was not anatomical but behavioral: animals apparently could not talk, and could not act rationally as opposed to automatically or instinctively. Yet to Descartes, the analogy between human and animal bodies was otherwise much closer than even anatomists had acknowledged; indeed it was nearly perfect. As he explained in the *Treatise on Man,* the "*machine de terre*" that constituted the body contained "only organs and mechanisms of such a type that you may well believe very similar ones to be present both in us and in many animals which lack reason as well." He referred specifics in this regard to "the anatomists."[29]

As many commentators have noted, Descartes's knowledge of brain anatomy, as displayed in his account of the pineal gland and its function, was imperfect. He accepted an essentially Galenic notion of the role of animal spirits in sensation and thought, identifying it in the *Treatise on Man* as "a certain very fine wind, or rather a very lively and pure flame," strained from the blood by arteries surrounding the pineal gland, which then distributed them.[30] As he made clear in the 1640s, first in letters and then in *Passions of the Soul*, the pineal gland was not only the seat of sensation but the "principal seat of the soul, and the place in which all our thoughts are formed."[31] Annie Bitbol-Hespériès comments, "Descartes designed the pineal gland or conarium as the point of corporeal articulation of the union of the soul to the body, and even 'to all the body'."[32] Although Galen had dismissed the notion of a central role for the pineal gland, and Vesalius had noted its presence in animals, Descartes persisted in his elevation of the gland for various reasons, including its singular, non-hemispherical nature and central position in the brain.[33]

The twofold challenge of Descartes's dualism and the animal-human anatomical analogy led to a number of responses. Marin Cureau de la Chambre made a

---

[27] Bynum 1973, 445.

[28] Descartes 1985, III:40; see also Clarke 2003.

[29] Descartes 1985, I:107.

[30] Descartes 1985, I:100.

[31] Descartes 1985, III: 143.

[32] Bitbol-Hespériès 1990, 97; my translation.

[33] Finger 1994, 26, 1995; Galen 1968, 419–420.

metaphysical argument for the essential equality of humans and animals not only in body but in mind, concluding that differences were a matter of degree. Thomas Willis, on the other hand, asserted that the structure of the brain as disclosed by dissection clearly showed that humans were unique. Stensen remained agnostic on this topic.

7.4 Cureau de la Chambre and Animal Cognition

Marin Cureau de la Chambre (1596–1669) received his medical training at Montpellier and gained the patronage of the royal chancellor Pierre Séguier in the 1630s mainly through his literary rather than his medical skills; he was a charter member of the Académie française in 1635. Among his talents were chiromancy and physiognomy – the discernment of character via the hands and face – and he employed these skills in the service of Séguier and, eventually, of the king. Like other court physicians, his therapeutic perspective inclined toward the chemical, although he wrote little on medical practice. He attended a number of salons and academies, including, in the 1650s, the academy of Henri-Louis Habert de Montmor. In 1666, Louis XIV's chief minister Jean-Baptiste Colbert chose him as one of the first members of the Paris Academy of Sciences, and several early anatomical works of the Academy made their first appearance in the form of pamphlets dedicated to him.[34]

Although he had been appointed to teach anatomy at the Jardin du roi upon its foundation in 1635, it is unlikely that Cureau ever performed a dissection. His interests in animals, and in what demarcated animals from humans, were metaphysical rather than anatomical, and in the 1640s and 50s he outlined a theory of animal soul and cognition that was quite different from that of Descartes. Pierre Bayle referred to Cureau as "one of the most illustrious Peripatetics of this century," but he admitted that Cureau's ideas were not conventionally Aristotelian, and they are in fact difficult to categorize.[35]

Cureau's *magnum opus* was his treatise on the passions, usually known as *Les charactères des passions* and published in five volumes between 1640 and 1662, with numerous reprintings (under various titles) in both France and the Netherlands. The first volume of this work outlined an even larger project that he referred to as "*l'art de connoistre les hommes*," the art of knowing men. This knowledge would penetrate into every aspect of what it meant to be human: the passions, the soul, vices and virtues, customs and morals, "the diverse inclinations of Men," the idea of beauty, the influence of climate, and not least, the relationship between humans and animals. In the preface to volume 1, he outlined five general rules of the art of knowing men: the first was "founded on the Characters of the Passions, the Virtues, & the

[34] Guerrini 2015.

[35] Bayle 1684, I:19–20.

Vices; and shows that those who have naturally the same Manner ["*Air*"] that accompanies the Passions or the Actions of the virtues & vices, are also naturally inclined to the same Passions & to the same Actions." The second "is taken from the Resemblance that Men have with Animals, & holds that those who have some part like those of the Others, have also the same inclinations as they." The third rule, founded on the "Beauty of the Sexes," held that men whose beauty was feminine were indeed effeminate in character, and women who had "something of a virile Beauty" shared "the inclinations of men." The fourth rule dictated a kind of climatic determinism with regard to human external appearance, while the fifth rule, "which is called Syllogistic," listed the particular signs by which one could discover, "by discourse and by reasoning," the *moeurs* or customs of various peoples. We may discern some Hippocratic influences here, but the main influence is indeed Aristotelian. The science of physiognomy depended heavily on Aristotle, who was among the earliest to argue that external appearances revealed inner character. These "inclinations" were innate and in turn directed the passions.[36]

Out of this mixture of topics came Cureau's very influential comments on animal passions and animal cognition. The passions, in Galenic medicine, were one of the six "non-naturals," defined as external as opposed to innate influences on health. The non-naturals acted on the humors and affected their balance within the body. Sometimes defined vaguely as "the affections of the mind," both Cureau and Descartes in his 1649 *Passions de l'âme* identified the passions more specifically with particular emotions such as love, hate, desire, and sadness that could directly influence the animal spirits. Cureau de la Chambre referred to the passions as "the actions of the virtues and vices," but also as "the movements of the Appetite, by which the soul *(l'âme)* tries to come near to good and avoid evil." The passions belonged specifically to the sensitive as opposed to the intellectual appetite, which governed the will. Therefore the passions were functions of the senses rather than of the mind.[37]

Although this looks like Cureau aimed toward a senses-versus-mind dualism, that is not the case. His discussions of the soul made it clear that he viewed both thinking and feeling, as well as life itself, as *conscious* acts: "one could say that all that is living knows, and all that knows is living."[38] His discussion of instinct and animal intelligence, first published in 1645, shed further light on this notion. Beginning with the chain of being and the idea of a hierarchy of intelligence, he argued that instincts were a form of innate knowledge.

Volume 2 of *Les charactères des passions* included "the courageous passions" such as loyalty and boldness, passions that animals as well as humans exhibited. He appended to the volume a lengthy discourse titled *Quelle est la connoissance des bestes, et iusques où elle peut aller* ("what is the knowledge of beasts, and how far

---

[36] Cureau de la Chambre 1658, "Advis necessaire au lecteur". All translations from Cureau de la Chambre are mine.

[37] Ibid.

[38] Cureau de la Chambre 1664, 217; see Wright 1991.

does it go") in which he directly addressed the question of animal intelligence. His use of the term "*bestes*" ("*bêtes*") or "*brutes*" rather than "*animaux*" was intended to draw a clear distinction between animals and humans; anatomists commonly referred to "*l'homme et les autres animaux*," but to "*bêtes*" when referring to animals alone.[39] Here Cureau reiterated the distinction he had earlier made between the sensitive and the rational soul. Nature was ordered by degrees, and thus the sensitive soul ranked below the rational. Yet the function of each was knowledge ("*la Connoissance*") or perhaps what we might call consciousness ("*connoissance*" is defined by the 1694 *Dictionnaire de l'Académie française* as an idea of something, as the power of judgment, and as the usage or function of the soul (*l'âme*)).[40] Cureau declared, "Because to feel, to conceive, to judge, to reason, all that is only to know."[41]

Cureau declared that both animals and humans held the power of perception. Because perception required an action of the soul, neither could receive sense impressions passively. Sensation, therefore, was a form of knowledge or consciousness. This in turn, argued Cureau, required the involvement of imagination and memory, traditionally considered to be actions of the conscious mind. He concluded that animals therefore possessed a degree of reason because they possessed a sensitive soul that included a degree of memory. If not the equal of humans, neither did they entirely lack consciousness. The fact that they could learn to do certain actions showed that. Cureau concluded that passions therefore emerged from consciousness; if animals exhibited passions such as anger or fear, then they must possess a kind of reason and even the ability to choose between good and evil, at least to some degree. Christian theology since Thomas Aquinas had considered the ability to choose as a hallmark of humanness.[42]

Cureau directly addressed the problem of speech as a criterion for animal consciousness in *Quelle est la connoissance des bestes.* "One would have to be extremely stupid," said Cureau, "not to notice that animals have voices," and that those voices differed according to the emotions they felt and expressed: "Pleasure or Pain, Hope or Fear." It was obvious that animals communicated with each other, and domesticated animals in addition communicated with humans. They communicated not only by voice but by gesture and facial expression – another indication of the centrality of physiognomy to Cureau's thought – and from his examples, he had spent considerable time observing dogs. Arguments against animal speech and cognition based on physiological differences between humans and animals were, he said, meaningless, and therefore so were arguments against animal cognition based on their supposed lack of speech.[43]

Although Cureau never mentioned Descartes in this work, it is hard to believe that when he wrote these words in 1645 he was not aware of Descartes's arguments

---

[39] See, for example, du Laurens 1600, Riolan 1628–29.

[40] *Dictionnaire* 1694.

[41] Cureau de la Chambre 1660b, 2; Wild 2008, 445.

[42] Cureau de la Chambre 1660b; Thomas 1983, 30–41.

[43] Cureau de la Chambre 1660b, 49–55.

about animal consciousness and speech in *Discourse on the Method,* published less than a decade earlier. This topic also appeared in the 1641 *Objections and Replies* to Descartes's *Meditations.* But Cureau readily acknowledged that even if animals possessed a kind of reason, they did not possess the "perfect and universal reason" of humans that assured their immortality.[44] In this he agreed with Descartes, who in a 1637 letter similarly argued that his views on animal consciousness supported the immortality of the human soul.[45]

7.5 Chanet's Critique and Cureau de la Chambre's Reply

Pierre Chanet, a physician of La Rochelle, responded in 1646 to Cureau 's *Quelle est la connoissance des bestes* in *De l'instinct et de la connoissance des animaux.* Chanet had penned a critique of the famous "*thériophiles*" Michel de Montaigne and Pierre Charron a few years earlier (*Considerations de la sagesse de Charron,* 1643), in which he argued that both humans and animals possessed instinct, which led us to eat when we were hungry or follow a path previously followed. But only humans had reason, which he defined as "a faculty of proving a thing by another different thing, and which proceeds by deliberation to a goal": in other words, the capacity to make logical inferences.[46] Chanet argued similarly in response to Cureau, adding that if humans and animals were essentially equal it would overturn the order of nature, and destroy that "*belle gradation,*" that beautiful hierarchy, that nature observes.[47] Chanet hastened to add that he did not attribute all animal actions to instinct; "they have as many external senses as we do; they also have memory, imagination, and a motive faculty."[48] But they did not have reason, which is "proper and specific to man." Employing a very Cartesian argument, Chanet claimed that one proof that beasts possessed reason would be if they could talk. But even though "their organs are not so different from ours," they could not speak "the language of Men." Parrots may "talk" but they do not therefore express thoughts.[49]

In response to Chanet, Cureau in 1648 published a more extended account of his views on animal intelligence, *Traité de la connoissance des animaux, où tout ce que a esté dict Pour, & Contre Le Raisonnement des Bestes, est examiné* (" treatise of the knowledge of animals, where all that has been said for and against the reasoning of animals is examined"). His dedication to his patron Séguier compared the judgement that the chancellor exercised over the men of the realm to the greater judgement exercised by God over all beings; "[God] does not contest the Reason of

44 Cureau de la Chambre 1660b, 19; Harrison 1998, 472–76.

45 Descartes 1985 III: 62–63.

46 Chanet 1643, 47, 57.

47 Chanet 1646, 2.

48 Ibid., 4.

49 Ibid., 120, 160–63.

Animals at all, & even believes that there would be some impiety to wish to suppress such a glorious mark of his Wisdom & of Divine Power."[50]

Like his contemporary Pierre Gassendi (who also frequented Montmor's salons in the 1650s), Cureau argued that the idea that humans needed to be distinguished from animals was specious; as we have seen, anatomists commonly referred to both as "*animaux*."[51] Also like Gassendi, he continued to distinguish two levels of reason: one based on perception of particulars that animals held, and a uniquely human higher reasoning based on the notion of universals. But unlike Gassendi, Cureau viewed this distinction as relative rather than absolute. He asserted that attributing animal actions to instinct rather than to a kind of reason (and free will) led to a theological as well as a philosophical problem. No one, he said, knew exactly how reason worked. He outlined three steps of perception:

> The First Understanding [*l'Entendement*] forms the image of things; then in the Second it unites or divides the Images; but when one comes to examine the Last in which consists Reasoning, one comes up short; and finds that the Discourse that like light makes the darkest things known, remains itself unknown & hides itself in the shadows.[52]

Descartes attributed this last function, the union of soul with body, to the pineal gland, but Cureau found no physical organ or site where sensation joined reason or understanding. He did not locate soul, or knowledge, or mind, in any specific part of the body.

Not only did attributing animal actions entirely to instinct imply that God had somehow left animals incomplete, but it also therefore required a kind of occasionalism – a direct intervention of God – to enable instinct. Creating animals with a degree of reason eliminated these problems.[53]

Cureau devoted a lengthy chapter in this lengthy book to the vexed question of animal speech, repeating his earlier arguments but adding material on the physiology of speech and the voice, on mutes, and on infants. If beasts have reason, he argued, they must be able to communicate – a neat reversal of Descartes's argument in *Discourse on the Method.* But nowhere did he mention Descartes, addressing his remarks entirely to Chanet. Cureau also drew no ethical conclusions about human behavior toward animals, nor did he address the issue of anatomical analogy, which he took for granted.[54]

A year later, Descartes's *Passions de l'âme* offered a completely different – and mechanical – explanation of the passions, the soul, and the mind. Descartes intended to eliminate the Aristotelian vegetative soul which governed bodily function in favor of mechanisms that performed these functions without conscious intervention. The passions are "motions of the [animal] soul, caused by the movement of the spirits." Therefore they are mechanically induced, but then, in humans, acted upon by the

---

[50] Cureau de la Chambre 1648, Epistre.

[51] Guerrini 2015.

[52] Cureau de la Chambre 1648, 4–5.

[53] Cureau de la Chambre 1648, 3–4. See also Boas 1933, 74–82; Harrison 1998, 475–479.

[54] Cureau de la Chambre 1648, 275–390.

rational soul, which is the seat of reason. Descartes admitted that some emotions, such as sadness, could be induced entirely by the rational soul without any external stimulus – you can think yourself into being sad, in other words, and experience it intellectually. But passions by definition depend on "impressions in the brain," and not on the soul. By this argument, then, animals were perfectly capable of experiencing the passions, but they did not have the rational capacity of humans to act upon them by moving the body in some way. Joy in an animal was pure emotion, without any rational goal or meaning.[55] Descartes believed that his mechanistic characterization of animal soul confirmed orthodox religious views of human exclusiveness and the immateriality of the soul and avoided the pitfalls of occasionalism. Indeed, in his dedication of *Principia philosophiae* (1644) to the Theology Faculty of the Sorbonne, he referred to himself as a "Christian philosopher."

Cureau accepted the analogy of human and animal bodies, and to some extent that of their minds or souls as well. He witnessed many of the early dissections and animal experiments of the Paris Academy and his ideas can be found particularly in the work of the man who organized these activities, Claude Perrault. In his *Essais de physique,* published in 1680, Perrault gave anatomical evidence for the mechanism of the human and animal body. Although Perrault was a mechanist, he retained the notion that "*connoissance*" was not only in the mind but pervaded a thinking and feeling animal body; among the examples he cited in his work on animal mechanism was a snake who lived in the garden behind the Academy. Even when its head and heart had been removed it somehow "knew" to hide behind its customary stone in the garden.[56]

7.6 Stensen and the Anatomy of the Brain

To the many who looked at animal and human function in the second half of the seventeenth century, dissection of the brain remained a difficult operation, and few pursued it with any depth. Among these few were Thomas Willis and Niels Stensen. While anatomical textbooks described the brain in greater or lesser detail, the descriptions had advanced little from those of Vesalius, and the illustrations continued to be borrowed from *De Fabrica.* Although these illustrations mainly depicted human brains – often *in situ* in the skull, with a face protruding disconcertingly below – anatomists such as Thomas Bartholin employed both animal and human brains in their descriptions. These works continued to take the analogy between human and animal for granted, but did not resolve its inherent contradictions.

Willis's first work on this topic, *Cerebri anatome*, appeared in 1664. A year later, Niels Stensen dissected a human brain in Paris in the spring of 1665, at the academy of the courtier Melchisédech Thévenot. In his description of this dissection,

[55] Descartes 1985 I: 356, 360–61.

[56] Guerrini 2015.

published in French 4 years later, Stensen agreed with his contemporaries on one point: "It is very certain that [the brain] is the principal organ of our soul."[57] But he agreed with them on very little else. Stensen asserted that the correct technique for dissection of the brain was much debated, and he admitted that even after numerous dissections he had not entirely penetrated the secrets of this organ (Stensen 1669, 8). Although he praised Willis's 1664 illustrations as the best of a bad lot, he disagreed with Willis's localization of brain functions, further described below.

Stensen reserved much of his critique for Descartes and his followers. Although the printer, Robert de Ninville, dedicated *Discours de Monsieur Stenon, sur L'anatomie du cerveau* to Cureau de la Chambre, Stensen's work bore little relation to Cureau's speculations on the mind. Taking Descartes in the recently published *Traité de l'homme* at his word – that he described only a mechanical model and not an actual human – Stensen found much to praise. "No one else has explained mechanically all the actions of man and particularly those of the brain …. Monsieur des Cartes speaks to us only of a machine, [but] nonetheless makes us see the insufficiency of what others teach us."[58] But the value of the *Traité de l'homme* remained on the level of inspiration, for anatomists such as Thomas Bartholin, Stensen's teacher in Copenhagen, and François de le Boë Sylvius, who had taught Stensen at Leiden, had already recognized its many anatomical errors, and Stensen was in addition unsympathetic to Cartesian systems. In a rather back-handed compliment, Stensen wrote "the excellence of his spirit [*esprit*] … in his Traité de l'Homme, covers the errors of his Hypotheses."[59]

Chief among Descartes's errors was the location and role of the pineal gland. Descartes implied, but did not directly claim, that the pineal gland was anatomically as well as functionally exclusive to humans. In fact, his correspondence makes clear that he had observed the gland in animals as well.[60] But the emphasis that Descartes placed on this gland as the source of animal spirits (which is strained from the vital spirits) gave it, as Stensen recognized, a central role in nervous function. Willis had published his account of the pineal gland in 1664 and noted its appearance in a number of animals. With painstaking anatomical detail, Stensen concluded that it was anatomically impossible that the pineal gland acted in the way Descartes had hypothesized. Perhaps most seriously, what Descartes had identified as arteries that carried vital spirit to the gland to be purified were in fact, said Stensen, veins that carried blood away from the gland toward the heart.[61]

Whether the pineal gland was exclusively human was not the issue to either Stensen or Willis. Certainly Stensen and his teachers Bartholin and Sylvius relied heavily on animal anatomy to talk about human structure and function, following

[57] Stensen 1669, 5.

[58] Ibid., 13.

[59] Ibid., 13–14.

[60] Finger 1995.

[61] Stensen 1669, 21.

the example not only of Harvey but of Jean Riolan.[62] Stensen singled out for praise the French anatomist Jean Pecquet, whose discovery of the thoracic duct in 1650 had taken many years of work – work, as Stensen well knew, that involved the dissection of over 100 living and dead animals.[63] In contrast, "the Anatomists, [who] open a thousand bodies in the Schools, only discover something by chance." Stensen went on, "To have some knowledge of the brain, it is necessary to dissect and examine as many heads, as there are species of animals, & different states in each species." He also recommended looking at fetal animals and experimenting on living ones: "There is also this reason to work on the brains of animals, that we [can] treat them as we wish."[64] Stensen argued that experiments could reveal "*actions animales*" common to all species, but he also acknowledged that the brains of animals such as birds and fish were very different from human ones; "and in the animals that approach most closely to us, I have not seen one, where I have not found some very obvious difference."[65] The analogy was not perfect, but it was good enough to give general information about brain function.

7.7 Willis and the Analogy of Nature

Richard Lower reported to Robert Boyle early in 1662 that he and Willis had begun to dissect human and animal brains. Among the animals they used were the usual cats, dogs, pigs, and sheep, but also horses, calves, goats, foxes, hares, geese, turkeys, fish, and a monkey.[66] Like many of his contemporaries, particularly Gassendi, Willis conceived of two kinds of soul (or mind); he called them the rational and the animal or corporeal soul, the "*anima brutorum*." Unlike Cureau de la Chambre, Willis sharply distinguished between these; only humans possessed a rational soul, which was immortal and the seat of reason and judgment.[67] He further divided the animal soul into vital and sensitive souls based respectively in the blood and nerves, echoing Galenic and Vesalian descriptions of vital and animal spirits that Descartes had also adopted.[68]

As is well known, Willis employed anatomical evidence to talk about localization of brain function. This was not the medieval organization in "ventricles," or even the more sophisticated localization that Galen had detailed in *Anatomical procedures*. In his 1667 work on the pathology of the brain, Willis argued that a number of illnesses could be attributed to physical disturbances of the brain and nervous

62 Guerrini 2013.

63 Stensen 1669, 36; Pecquet 1651.

64 Stensen 1669, 54.

65 Ibid., 55.

66 Frank 1980, 126.

67 Willis 1672; Frank 1980, 67, 68, 130–31; Dewhurst 1968.

68 Willis 1672; Bynum 1973, 449.

system.[69] I won't go into the details of the atomistic chemical composition of the animal spirits that underlay his conclusions. What I wish to emphasize is Willis's anatomical differentiation of the parts and functions of the brain (and therefore of the nervous system), and how much he based this on animal evidence.

From the outset of his work on the anatomy of the brain, Willis accepted a close analogy, even an identity, between human and animal brains.[70] "That the perfect knowledge of the Brain and its parts may be gained, it is necessary not only to dissect and look into mens Heads, but all other kind of living Creatures heads." Human heads were not always readily available, and what he referred to as their "immense bulk," while a convincing argument for superior intelligence, was a hindrance in seeking specific structures via dissection.[71] The "form and composition" of the brains of quadrupeds, he wrote, "were little different" than that of humans, except that they were smaller; therefore "I was the more satisfied to compose a certain Anatomy of the Brain by the frequent dissection of all sorts of living Creatures." "Certainly from such a compared Anatomy," he continued, "not only the faculties and uses of every Organ, but the impressions, influences, and secret ways of working of the sensitive Soul it self will be discovered."[72] Unlike Stensen, who maintained that there were fundamental differences between animal and human brains, Willis therefore argued that animals were actually better subjects for a thorough anatomy of the brain than humans.

As William Bynum detailed 40 years ago, Willis, like all of his contemporaries, employed anatomical methods to make physiological claims. Following Harvey's methods of dissection of living and dead animals across a number of species, Willis made claims about how the body worked from how it was constructed. Where Cureau de la Chambre, followed by Perrault, worked from behavior to structure, Willis worked from structure to behavior. Vesalius had expressed skepticism that the folds of the cerebrum had any meaningful relationship to human intelligence, pointing out, with Galen, that animals too had these "convolutions." Vesalius admitted that human brains had deeper convolutions, which might provide the human brain with more blood and therefore more nourishment; but he did not then connect that circumstance to human intelligence, agreeing with Galen that the quantity of animal spirits rather than the morphology of the brain determined intelligence.[73]

For Willis, on the other hand, these convolutions were directly related to the "dispensation" of the animal spirits and therefore to imagination and intelligence, specifically human characteristics. The more space there was for the spirits to act, the more opportunity for "the various Acts of Imagination and Memory" to be imprinted in the brain. "Hence these folds or rollings about … are far more and greater in a man than in any other living Creature, to wit, for the various and

[69] Willis 1667.

[70] See also Crignon 2017.

[71] Willis 1684, 46.

[72] Willis 1684, 46.

[73] Vesalius 1952.

manifold actings of the superior Faculties."[74] To be sure, Willis undoubtedly looked at many more human and animal brains than either Galen or Vesalius. The size of the human brain, while a disadvantage from the point of view of the anatomist, provided a distinct proof of superior human imagination and memory; a cat, who remembered only "what the Instincts and needs of Nature suggest," had necessarily a smaller brain with fewer folds.[75] Although form dictated function to Willis, in this case at least teleology dictated form. Quite apart from its anatomical connections to the nervous system, the cerebrum had to be the seat of imagination and memory because, according to Willis's observations, it was so much more complex, as well as larger, in humans. The brain was the seat of "wit" and passion.[76]

The cerebellum, on the other hand, varied much less among animals and humans. Rejecting the ancient notion that it was simply "another Brain," Willis "diligently and seriously meditated on the office of the Cerebel … at length, from the Analogy and frequent Ratiocination," he concluded that it was a "peculiar Fountain of Animal spirits … wholly distinct from the Brain." He believed he had shown anatomically that animal spirits flow from the cerebrum "into the nervous stock" by which all conscious motion occurs. The animal spirits flowing from the cerebellum, however, activated a different set of nerves; its functions were those common to humans and animals, such as the involuntary motion of the heart and lungs. Such straightforward actions, Willis argued, did not require the convolutions necessary for deep memories or imagination, but conformed to the smoother and more "orderly" structure of the cerebellum. This argument echoed Descartes's structural argument for the pineal gland being the seat of the soul because of its shape and location. And indeed, once he began to look for them, Willis found the connections between the cerebellum and the nerves upon which involuntary motion depended.[77]

Thus far, then, Willis offered a concept of the brain that differed only in degree from animal to human. Humans had superior memory and imagination to animals, but animals did not entirely lack these attributes. Thus far, too, Willis's concept of nervous function was strictly material. Animal spirits constituted an actual fluid, and acted on material nerves. Willis did not go on to trace the localization of nervous (as opposed to brain) function, because he had neither the means or the knowledge to do so; he was not Magendie.[78] His arguments for the localization of brain function showed that he did not accept Cureau de la Chambre's notion of a sensitive soul diffused throughout the body. It is not clear that he knew of this theory; most of Cureau's worked remained in French, with the important exception of the *Traité de la connoissance des animaux*, which had been translated into English in 1657. Willis also would not have agreed with Cureau that possession of a sensitive soul – of a degree of memory and imagination –

[74] Willis 1684, 76.

[75] Ibid.

[76] Ibid., 100–101; Martensen 2004, 138–39.

[77] Willis 1684, 91–92.

[78] Frank 1980.

equated to a degree of rationality. Although Willis believed that memory and imagination were necessary for reason, they were not sufficient.

Cureau had readily acknowledged that the rational soul, the soul that only humans possessed, was immaterial. He made no attempt to locate it in the brain, and to us its function seems mainly theological: the differences between humans and animals were of degree rather than kind, except for the possibility of salvation; the "perfect and universal reason" that humans possessed allowed them to know God. It was the sensitive or animal soul that carried the principle of life, as Perrault's snake had shown. Willis too argued that the rational soul was immaterial, but his anatomical framework seated both the sensitive and the rational souls in the brain; he identified the "Court of the Phantasie" as the seat of the rational soul.[79] Elsewhere he appeared to identify the corpus callosum – the broad band of nerve fibers joining the two hemispheres of the brain – as the "region of the Brain where the principle faculties of the knowing or understanding soul reside," but this identification is not entirely clear.[80]

Rina Knoeff has detailed the role Willis assigned to the intercostal nerves in bridging the brain, heart, and lower viscera, conveying the animal spirits to the lower body, and mediating and regulating the actions of the passions. According to Willis, these nerves were anatomically different in animals and humans – humans had an extra connection between the brain and the heart – and they served this additional regulatory function only in humans; this function did not exist in animals. As Knoeff points out, this arrangement "kept the rational soul beyond the reach of the anatomical knife."[81] But, as she notes, adding another mechanism to distinguish humans from animals did not address the relationship between the immaterial soul and the material body, and moreover left unresolved the distinction between the rational and sensitive souls.

However, Willis's material grounding of the rational and sensitive souls allowed him to talk about nervous diseases and their therapies in ways that neither Stensen nor Cureau did. Although Cureau was a royal physician, he talked little about medicine in his writings, and Stensen admitted that his interests lay in anatomy rather than therapy. Willis identified a number of diseases that were based in the brain and nervous system, and offered specific therapeutics based on his chemical concept of the blood and animal spirits. In addition, he offered mechanical explanations of "stupidity" and "foolishness." But, as with the role of the intracostal nerves, the close relationship between the rational and sensitive souls and their functions also served to blur the boundaries between them: stupidity, although seemingly a failure of reason, was purely a result of the texture of the cerebrum and the quality of the animal spirits.[82] Willis could therefore give a good account of how brain injuries could cause "foolishness," or why certain individuals had less wit than others. But

[79] Willis 1683, 41–42.

[80] Ibid., 153; Bynum 1973, 456.

[81] Knoeff 2004, 431.

[82] Willis 1683, 209–14; Cranefield 1961.

he ended up by giving the rational soul very little to do, other than, like Cureau, to ensure salvation.

7.8 Conclusion

The overall impact of these ideas on mental and physical health, therefore, was mixed. Only Willis viewed himself primarily as a practicing physician, and left a number of works on therapeutics. Nonetheless, his emphasis on anatomy, dissection, and anatomical methods led him to rely heavily on animal bodies as evidence of human function, including human mental function. Although he fervently believed in the existence of an immaterial rational soul that was unique to humans, by locating it in the brain – by acknowledging "brainhood" – he inevitably elided its functions with that of the material animal soul. According to his materialist concept of illness, mental illness as well as physical illness owed to the material conditions of the animal soul and the physical conditions of the brain.

The analogy of animals to humans had therefore become a continuum without a distinct breaking point, and subsequent work only narrowed what little remained of that divide. The work of John Ray and Edward Tyson on the brain of the porpoise and Tyson on the chimpanzee further obscured the physical distinctions between human and animal brains. In 1680, Ray and Tyson noted the large size of the porpoise's brain; in Tyson's words, "Mr. Ray saith that the largeness of the Brain of the fish, and the Correspondence of it to that of a Man's, argues this Creature to be of a more than Ordinary Wit and Capacity."[83] In his preface to *Phocaena, or the Anatomy of a Porpess*, Tyson (as had Perrault a few years earlier) argued that animal anatomy could give insight into human diseases.

Only when he discussed the anatomy of the brain, toward the end of his treatise, did Tyson actually address any human comparisons, citing Willis's illustration of and description of the human brain. But as Lawrence Kruger has recently argued, Tyson evaded the consequences of the similarities of human and animal brains by concluding in his 1699 work on the anatomy of the chimpanzee that God gave humans an immaterial soul, and this rather than the brain was responsible for human intelligence. The "pygmie's" brain was indeed so similar to the human that "you may justly suspect I am describing that of a Man." But he quickly stepped back from asserting any closer identity.[84] This seems a retreat from Willis's claim; Tyson is clear that "if all depended on the *Organ*, not only our *Pygmie*, but other *Brutes* likewise, would be too near akin to us."[85] Willis established, seemingly beyond dispute, the close relationship between mental illness and the physical body, largely by looking at animal brains. But Cureau de la Chambre had balked at attributing full

[83] Tyson 1680, 40.

[84] Tyson 1699, 54–55; Kruger 2003, 345–46.

[85] Tyson 1699, 55.

consciousness to animals, despite the logical implications of his argument, and Tyson pulled back from Willis's assertion that the rational soul resided in the brain. Although the analogy between human and animal bodies remained and even grew stronger, "brainhood" had to wait to be fully established.

Bibliography

Anatomia Cophonis. 1927. In *Anatomical texts of the earlier middle ages: A study in the transmission of culture with a revised Latin text of the Anatomia Cophonis and translations of four texts*, ed. G.W. Corner. Washington, DC: Carnegie Institution.

Aristotle. 1965. *History of animals*. Books 1–3, Trans. A.J. Peck. Loeb Classical Library. Cambridge, MA: Harvard University Press.

Atran, Scott. 1987. Origin of the species and genus concepts: An anthropological perspective. *Journal of the History of Biology* 20: 195–279.

Balme, D.M. 1972. *Aristotle's De partibus animalium I and De generatione animalium I*. Oxford: Clarendon Press.

Bayle, Pierre. 1684. *Nouvelles de la république des lettres*, mars 1684.

Bitbol-Hespériès, Annie. 1990. *Le principe de vie chez Descartes*. Paris: Vrin.

Boas, George. 1933. *The happy beast in French thought of the seventeenth century*. Baltimore: Johns Hopkins University Press.

Bynum, William F. 1973. The anatomical method, natural theology, and the functions of the brain. *Isis* 64: 444–468.

Chanet, Pierre. 1643. *Considerations de la sagesse de Charron*. Paris: Claude Le Groult and Jean Le Mire.

———. 1646. *De l'instinct et de la connoissance des animaux*. Toussaincts De Gouy: La Rochelle.

Clarke, Desmond. 2003. *Descartes's theory of mind*. Oxford: Clarendon Press.

Cranefield, Paul. 1961. A seventeenth-century view of mental deficiency and schizophrenia: Thomas Willis on 'stupidity or foolishness'. *Bulletin of the History of Medicine* 35: 291–316.

Crignon, Claire. 2017. How animals may help us understand men. Thomas Willis's *anatomy of the brain* (1664) & *Two discourses concerning the Soules of Brutes* (1672). In *Human & animal cognition in early modern philosophy & medicine*, ed. Stefanie Buchenau and Roberto Lo Presti, 173–185. Pittsburgh: University of Pittsburgh Press.

Cureau de la Chambre, Marin. 1648. *Traité de la connoissance des animaux*. Paris: Pierre Rocolet.

———. 1658. *Les charactères des passions*. Vol. 1. Amsterdam: Antoine Michel.

———. 1660a. *Les charactères des passions*. Vol. 2. Paris: Pierre Rocolet.

———. 1660b. *Quelle est la connoissance des bestes, et iusques où elle peut aller.* In *Les charactères des passions*. Vol. 2, separately paginated.

———. 1664. *Le système de l'âme*. Paris: Jacques Allin.

Descartes, René. 1985. *Philosophical works of Descartes*, ed. John Cottingham, Robert Stoothoff, and Dugald Murdoch, 3 vols. Cambridge: Cambridge University Press.

Dewhurst, Kenneth. 1968. Willis and Steno. In *Steno and brain research in the seventeenth century*, ed. Gustav Scherz, 43–48. Oxford: Pergamon Press.

Dictionnaire de l'académie Française. 1694. https://artfl-project.uchicago.edu/content/dictionnaires-dautrefois. Accessed 30 Apr 2014.

Dolan, Brian. 2007. Soul searching: A brief history of the mind/body debate in the neurosciences. *Neurosurgical Focus* 23 (2007): 1–7.

du Laurens, André. 1600. *Historia anatomica humani corporis*. Paris: Marc Orry.

Finger, Stanley. 1994. *Origins of neuroscience*. New York: Oxford University Press.

———. 1995. Descartes and the pineal gland in animals: A frequent misinterpretation. *Journal of the History of the Neurosciences* 4: 166–182.

Frank, Robert G. 1980. *Harvey and the Oxford physiologists*. Berkeley: University of California Press.
Galen. 1956. *On anatomical procedures*. Trans. Charles Singer. Oxford: Oxford University Press.
———. 1962. *On anatomical procedures, the later books*. Trans. M.C. Duckworth, ed. M.C. Lyons and B. Towers. Cambridge: Cambridge University Press.
———. 1968. *On the usefulness of the parts of the body*. Trans. Margaret Talmadge May. Ithaca: Cornell University Press.
Guerrini, Anita. 2013. Experiments, causation, and the uses of vivisection in the first half of the seventeenth century. *Journal of the History of Biology* 46: 227–254.
———. 2015. *The courtiers' anatomists. Animals and humans in Louis XIV's Paris*. Chicago: University of Chicago Press.
Hankinson, R.J. 1991. Galen's anatomy of the soul. *Phronesis* 36: 197–233.
Harrison, Peter. 1998. The virtues of animals in seventeenth-century thought. *Journal of the History of Ideas* 59: 463–484.
Hippocrates. 1978. The Sacred Disease. In *Hippocratic writings*, ed. G.E.R. Lloyd. London: Penguin.
Knoeff, Rina. 2004. The reins of the soul: The centrality of the intercostal nerves to the neurology of Thomas Willis and in Samuel Parker's theology. *Journal of the History of Medicine and Allied Sciences* 59: 413–440.
Kruger, Lawrence. 2003. Edward Tyson's 1680 account of the 'Porpess': Brain and its place in the history of comparative neurology. *Journal of the History of the Neurosciences* 12: 339–349.
Lloyd, G.E.R. 1961. The development of Aristotle's theory of the classification of animals. *Phronesis* 7: 59–81.
———. 1975. Alcmaeon and the early history of dissection. *Sudhoffs Archiv* 59: 113–147.
Martensen, Robert L. 2004. *The brain takes shape*. Oxford: Oxford University Press.
Pranghofer, Sebastian. 2009. 'It could be seen more clearly in unreasonable animals than in humans': The representation of the rete mirabile in early modern anatomy. *Medical History* 53: 561–586.
Riolan, Jean. 1628–29. *Les oeuvres anatomiques*. Paris: Denys Moreau.
Sorabji, Richard. 1993. *Animal minds and human morals*. Ithaca: Cornell University Press.
Stensen, Niels. 1669. *Discours de Monsieur Stenon, sur L'anatomie du cerveau*. Paris: Robert de Ninville.
Thomas, Keith. 1983. *Man and the natural world*. New York: Pantheon.
Tyson, Edward. 1680. *Phocaena, or the anatomy of a Porpess*. London: Benj. Tooke.
Vesalius. 1952. *Vesalius on the human brain*. Intro. and Trans. Charles Singer. London: Oxford University Press.
Vidal, Fernando. 2009. Brainhood, anthropological figure of modernity. *History of the Human Sciences* 22: 5–36.
Von Staden, Heinrich. 1989. *Herophilus the art of medicine in early Alexandria*. Cambridge: Cambridge University Press.
———. 2013. Writing the animal: Aristotle, Pliny the Elder, Galen. In *Writing science: Medical and mathematical authorship in ancient Greece*, ed. Markus Asper and Anna Maria Kanthak, 111–144. Berlin: Walter de Gruyter.
Wild, Markus. 2008. Marin Cureau de la Chambre on the natural cognition of the vegetative soul: An early modern theory of instinct. *Vivarium* 46: 443–461.
Willis, Thomas. 1683. *Two discourses concerning the Soul of Brutes*. Trans. Samuel Pordage. London: Thomas Dring et al.
———. 1684. The anatomy of the brain. In *Dr Willis's practice of physick*. London: for T. Dring et al., separately paginated.
Wilson, Malcolm. 1997. Analogy in Aristotle's biology. *Ancient Philosophy* 17: 335–358.
Wright, John. 1991. The embodied soul in seventeenth-century French medicine. *Canadian Bulletin of Medical History* 8: 21–42.

Chapter 8
Lockean Self-Diagnosis

Peter R. Anstey

Abstract This chapter examines John Locke's conceptions of the soul, the understanding and the self in order to shed light on his preoccupation with his own mental and bodily wellbeing. A practitioner of both philosophy and physic, Locke provides a fascinating case study of early modern self-diagnosis. His rich and innovative account of the self, with its focus on the moral self, the conscience and having a property in one's own person, provide the backdrop from which we can understand his concerns and prescriptions for the health of the body and the mind.

Keywords Conscience · Consciousness · Health · Moral man · Person · Personal identity · Self · Soul · Understanding · Wellbeing

Abbreviations

Corr.	Locke, John. *The Correspondence of John Locke*, 8 vols, ed. Esmond S. de Beer. Oxford: Clarendon Press, 1976–1989.
E	Locke, John. *An Essay concerning Human Understanding*, ed. Peter H. Nidditch. Oxford: Clarendon Press, 1975.
Locke Works	Locke, John. *The Works of John Locke*, 10 vols. London: Tegg, 1823.
STCE	Locke, John. *Some Thoughts Concerning Education*, eds John W. Yolton and Jean S. Yolton. Oxford: Clarendon Press, 1989.

Locke was both a philosopher and a physician. He was therefore well qualified to undertake quite different forms of self-diagnosis. On the one hand he 'diagnosed' the nature of the self as a philosophical category and its implications for mental

P. R. Anstey (✉)
The University of Sydney, Sydney, Australia
e-mail: peter.anstey@sydney.edu.au

G. Manning (ed.), *Testimonies: States of Mind and States of the Body in the Early Modern Period*, Archimedes 57,
https://doi.org/10.1007/978-3-030-39375-5_8

health. On the other hand, he self-diagnosed his own medical conditions, the state of his own body. Furthermore, he was acutely aware that the two are intimately related: that one's bodily health is tied to one's mental health. He wrote to Mary Clarke in 1696:

> you will often reflect upon what you say in your letter to Mr. Clarke [her husband], that you find your health as the quicksilver in the weatherglass rise and fall as your temper of mind is. It is that which I know to be so, and do tell you but that half your cure depends on the Doctor's prescriptions, the other half is in your own mind. (*Corr.*, 5: 701)

These two interlocking facets of human nature—bodily health and mental health—are things about which Locke had a deep and abiding interest.

In fact, some 12 years earlier Locke had written a series of letters to Mary's husband Edward Clarke giving advice on the education of their son. Locke's advice began with Juvenal's saying: *Mens sana in corpore sano* (Sound in mind, sound in body) which he describes as 'a short but full description of the most desirable state we are capable of in this life' (*Corr.*, 2: 626; see also *SCTE*, 83). And this is carried over into the opening lines of *Some Thoughts Concerning Education* which was a substantially augmented version of his correspondence with Clarke. Interestingly, this relation between the sound body and the sound mind is the organising principle for the first 33 sections of *Some Thoughts Concerning Education.* Locke takes up the body first and offers prescriptions for the body of a young boy in sections 2–30 and then transitions to the mind in section 31:

> §31. Due care being had to keep the Body in Strength and Vigor, so that it may be able to obey and execute the Orders of the *Mind*; The next and principal Business is, to set the *Mind* right, that on all Occasions it may be disposed to consent to nothing, but what may be suitable to the Dignity and Excellency of a rational Creature. (*STCE*, 102–103)

Yet *Some Thoughts Concerning Education* is concerned with the formation of the bodies and minds of children, whereas our concern here is with Locke's own experience of and prescriptions for his body and mind, that is, with Locke's *self*-diagnosis.

There are a number of ways of approaching this topic. One could, for example, survey all of the evidence about Locke's own health: his medical self-diagnoses, self-administered therapies and regimens, and the advice that he sought from his friends. Again, one could, as Sorana Corneanu has recently done (Corneanu 2011, esp. chap. 5), examine Locke's views on the cure of the mind. Where does Locke fit in in the *cultura animi* tradition and what implications are there for bodily health arising from his views on the cure and cultivation of the understanding? Alternatively, one could examine Locke's philosophy of the self and follow it through to its implications for the cultivation of health and, beyond that, to determining our body's rightful place in nature. It is this third approach that I take here, fully conscious that it is only one route in to a very fertile terrain. I want to sketch Locke's theory of the self and then to explore its implications for what one can know about nature and how one should live in the world.

Before delving into Locke's philosophy of the self, however, it is important to point out that Locke's own intellectual formation was tied to contingencies that arose from his preoccupation with his own health. It is always dangerous to deal in biographical counterfactuals. However, it is almost certain that had Locke had better health or less concern for his own wellbeing, he would have become a different philosopher, or at least he would have had vastly different intellectual interactions and influences.

We get a glimpse of this in a pair of back-to-back letters he wrote from Salisbury, where he had gone to recover his health, to Dr. John Mapletoft in October 1671. He explains how he had first removed to Oxford where 'I mended apace, and my cough sensibly abated'. But on moving to Salisbury he said, 'I feare the aire will not be soe advantageous to me … I have but made a stand, if not gone backwards, in this watry place' (*Corr.*, 1: 360). On the verso side he claims in a second letter, 'I must conclude my carcasse to be made of a very ill composition that will not grow into good plight in fresh aire' and 'whilst my minde is at perfect ease … methinkes my body should batten' (*Corr.*, 1: 361).

Locke's concerns with his own health were so acute that it was probably the primary motivation for his leaving London and undertaking an extended trip to France in November 1675. His own health is a recurring theme in his letters from France over the next few years and his personification of health as his mistress is telling. For, after recovering from severe concussion and a tertian ague he wrote to Mapletoft, who had recently become engaged, in the following words:

> My health, which you are so kinde to in your wishes, is the only mistris I have a long time courted, and is soe coy a one that I thinke it will take up the remainder of my days to obteyn her good graces and keep her in good humor. She hath of late been very wayward, but I hope is now comeing about again. I should be glad that my constant addresses should at last prevaile with her, that I might be in a better condition to enjoy and serve you. (*Corr.*, 1: 491)[1]

8.1 Locke's Distinctions Between Soul, Understanding and Self

In order to understand Locke's diagnosis of the self, it is important to set out his basic terms of reference, for they are substantially different from those that we use today. Locke's view is founded upon three closely related notions: the soul, the understanding and the self. There are some very important things to note about this triad of notions.

The first is the absence from this list of the mind. Locke's *Essay* is an essay concerning human understanding or *intellectus*. It is not an essay concerning the human mind. Indeed, it is slightly jarring to speak of the philosophy of mind in seventeenth-century England: the category only really came in to philosophy in the mid-eighteenth century. Of course, Locke is always using the term 'mind': we have encountered it already in the quotations with which this chapter began. The problem, however, is that he uses it as a synonym both for 'soul', for the combination of

[1] For Locke's self-diagnosis and treatment for his tertian ague in his journal, see Dewhurst 1963, 80–82.

the understanding and the will, and often simply for the understanding. It is this semantic breadth that renders the term 'mind' the least interesting item in his vocabulary about the self.[2]

Second, the understanding is one of the faculties of the soul. It is complemented by and intimately related to the will. In spite of the fact that the *Essay* concerns the understanding and not the will, Locke also presents a theory of the will and its relation to the understanding in Book Two, chapter 21. This theory underwent very significant changes as Locke revised the *Essay*, so much so that one can only adequately expound each stage of his view by referring to the specific edition numbers in which it appears.[3] For better or worse, I will not be referring to Locke's view of the will here.

Third, returning to the triad of notions, for Locke the soul is that substance which has the power of thought. In the *Essay* and in the Stillingfleet correspondence, the question of the ontological status of the soul looms large. Is it a material or an immaterial substance? Locke can see no reason why the former should be ruled out: the soul might be material (*E* IV. iii. 6). In his theological writings of the last 10 years of his life, however, Locke's interests shift to a different conception of the soul: one in which the question of its ontological status is beside the point.[4] Happily, not much hangs on the question of the ontological status of the soul for our concern with the nature of the self and our knowledge of the nature of things. Of course, much does hang on the ontological status of the soul for Locke's view of the Resurrection and life after death, and these matters were of great importance to him,[5] but we will ignore them here.

Fourth, Locke's preferred term for the self is 'person' and I will use the two terms 'self' and 'person' interchangeably, as he does.[6] So Locke has the soul, which is a thinking substance, the understanding which is a faculty of the soul, and the self.

8.2 Locke's Distinctions Between Masses of Matter, Man and Self

We are not yet ready to discuss Locke's diagnosis of the self, for, we need to spell out another set of distinctions. In Book Two, chapter 27 of the second edition of the *Essay*, where Locke sets out his theory of the self, he gives a threefold distinction between masses of matter, man and person. His initial motivation for the chapter is to specify principles of individuation for different kinds of things. With regard to mere material bodies, Locke has in mind aggregates of atoms or corpuscles. He

[2] For further discussion, see Anstey 2015, 221–225.

[3] For a thorough treatment, see Stuart 2013, chaps 9 and 10.

[4] See Anstey 2015, 229–234.

[5] See, for example, *Second Reply to the Bishop of Worcester*, *Locke Works*, 4: 301–334.

[6] See Weinberg 2016, 147 n. 7.

claims that if one atom is removed or replaced in such a cluster, it is a different thing (*E* II. xxvii. 3). This view is sometimes called Mereological Essentialism because each part of a particular mass of stuff is essential to it being that particular mass of stuff.

With regard to animals, including men, and plants, Locke claims that it is the arrangement of parts giving rise to functional systems that constitutes the principle of individuation and identity over time. Even if every atom of an oak tree were to be replaced over time, it is the arrangement of parts and the persistence of the functional systems that result, that constitutes the principle of individuation and therefore of identity over time (*E* II. xxvii. 3–6). Thirdly, with regard to persons or the self, it is continuity of consciousness that constitutes both the principle of individuation and of identity over time. Here is how Locke puts it:

> [A person] is a thinking intelligent Being, that has reason and reflection, and can consider it self as it self, the same thinking thing in different times and places; which it does only by that consciousness, which is inseparable from thinking, and as it seems to me essential to it. (*E* II. xxvii. 9)

From this Locke derives a criterion of psychological continuity for the identity of persons over time:

> For since consciousness always accompanies thinking, and 'tis that, that makes every one to be, what he calls *self*; and thereby distinguishes himself from all other thinking things, in this alone consists *personal Identity*, *i.e.* the sameness of a rational Being: And as far as this consciousness can be extended backwards to any past Action or Thought, so far reaches the Identity of that *Person*. (Ibid.)

Just what Locke means by continuity of consciousness is a vexed issue in Locke scholarship.[7] In my view, the core idea is that a person is a collection, or better, an equivalence class, of conscious thoughts. Each thought, be it a memory or state of current conscious thought, is the object of a perception that it is part of the same thinking thing that has existed in different times and places. So, just as I can be conscious that a particular thought derives from memory or directly from sensation, so I am conscious that my particular memory is part of my thinking self.

Three points are worth making about Locke's view of consciousness. First, from a contemporary perspective we would stipulate that this consciousness is also a Same-Order-Perception rather than a Higher-Order-Perception, on pain of an infinite regress.[8] Thus, for Locke consciousness cannot be a perception of a perception, because if it were there would be no end to us being conscious of being conscious. Instead perception is best thought of as a constituent of the act of perception itself, what Shelley Weinberg helpfully describes as an 'awareness internal to ordinary perception' (Weinberg 2016, 32). But this is to go beyond anything that Locke says. Second, Locke claims, famously, in *Essay* II. xxvii that the self extends temporally as far as consciousness of one's memories extends: personal identity has a diachronic as well as a synchronic dimension.

---

[7] The best survey and analysis is in Thiel 2011, Parts II and III.

[8] For further discussion, see Coventry and Kriegel 2008.

Third, and this is often overlooked, on one occasion he claims that consciousness is also *spatially* extended:

> *Self* is that conscious thinking thing … and so is concern'd for it *self*, as far as that consciousness extends. Thus every one finds, that whilst comprehended under that consciousness, the little Finger is as much a part of it *self*, as what is most so. Upon separation of this little Finger, should this consciousness go along with the little Finger, and leave the rest of the Body, 'tis evident the little Finger would be the *Person*, the *same Person*; and *self* then would have nothing to do with the rest of the body. (*E* II. xxvii. 17; see also §18)

It is clear from this passage that Locke views what we now call bodily sensations and proprioception as forms of self-consciousness. And taken literally, it might seem that Locke is committed to the view that to be conscious of where my little finger is or that it is in pain entails that my little finger is conscious. Now, the case of phantom limbs poses a problem for this view and Locke had long been aware of the phenomenon. He wrote to Robert Boyle in 1667 of a young woman amputee

> who some years since had the misfortune to have one of her legs cut off, which yet very frequently does so pain her, that I have heard her more than once grievously complain of vehement torture, sometimes in the heel, sometimes in the instep, and at other times in the toes of her dead and buried foot. (*Corr.*, 1: 315)

Perhaps Locke would have claimed that we have here a case of sensory illusion, akin to the modern Müller-Lyer illusion: 'sometimes being wakened by the pain, she has (before recollection) put her hand down to feel the grieved part, as if her leg had not been cut off' (ibid.). Sadly, Locke did not discuss this phenomenon in relation to consciousness. However, Locke's point in the 'little finger passage' is merely that the person is where consciousness is extended, whether in a finger alone or the whole human body.

Turning back to Locke's account of the self, one advantage of his view is that it avoids any commitments as to the ontological status of persons, and Locke seems to have had no interest in the question. (This has not deterred his interpreters from attempts to answer this question for him.[9]) Another advantage is that it gives him, not simply a principle of individuation, but also a criterion of identity over time. Having an account of personal identity over time, in turn, allows him an account of moral responsibility and Locke goes on to stress that 'person' or 'self' is a forensic term.

> *Person*, as I take it, is the name for this *self*. Where-ever a Man finds, what he calls *himself*, there I think another may say is the same *Person*. It is a Forensick Term appropriating Actions and their Merit; and so belongs only to intelligent Agents capable of a Law, and Happiness and Misery. This personality extends it *self* beyond present Existence to what is past, only by consciousness, whereby it becomes concerned and accountable, owns and imputes to it *self* past Actions, just upon the same ground, and for the same reason, that it does the present. (*E* II. xxvii. 26)

It seems natural, therefore to view *self*-diagnosis as involving the owning and imputing to oneself past actions. It also involves the moral evaluation of past and present

[9] See, for example, Rickless 2015 and LoLordo 2012 and for criticisms of LoLordo, Anstey 2013b.

actions that one imputes to oneself. The relevant faculty here would seem to be the conscience. Earlier in Book One Locke defines conscience as 'nothing else, but our own Opinion or Judgment of the Moral Rectitude or Pravity of our own Actions' (*E* I. iii. 8). Locke never fully develops his conception of the conscience beyond claiming that it is that faculty by which we accuse or absolve ourselves. It is that faculty by which we perform acts of first person moral evaluation. The term 'conscience' is most common in his writings on toleration, though in that context it is primarily associated with the notion of liberty of conscience and the absence of coercion in religious belief and practice (*Locke Works*, vol. 6). What is implicit is that the exercise of conscience involves the ethics of belief that Locke develops in the *Essay* and *Of the Conduct of the Understanding*.

Issues pertaining to the moral evaluation and appropriation of past actions bring out the uniqueness of Locke's position and its potential strengths and weaknesses. On the one hand, Locke's account of the self allows for the loss of memories and both the development and fragmentation of the self. This accords well with many of our basic intuitions about personhood, some of which are reflected in common expressions such as 'she was beside herself' or 'she's not her old self'.[10] It also harmonises well with our experience of illness and cognitive ageing, particularly dementia. Indeed, reflection upon this provides Locke with another case of the way the mind is dependent upon the body:

> that the Constitution of the Body does sometimes influence the Memory; since we oftentimes find a Disease quite strip the Mind of all its *Ideas*, and the flames of a Fever, in a few days, calcine all those Images to dust and confusion, which seem'd to be as lasting, as if graved in Marble. (*E* II. x. 5)

Earlier in the same section he laments the failing of memory:

> The Memory in some Men, 'tis true, is very tenacious, even to a Miracle: But yet there seems to be a constant decay of all our *Ideas*, even of those which are struck deepest, and in Minds the most retentive; so that … the Print wears out, and at last there remains nothing to be seen. …our Minds represent to us those Tombs, to which we are approaching; where though the Brass and Marble remain, yet the Inscriptions are effaced by time, and the Imagery moulders away. (Ibid.)

On the other hand, for all its intuitive appeal, some find Locke's theory objectionable in so far as it stipulates equivalence rather than strict identity as the salient relation between person stages. Thus, one classic objection of Thomas Reid is that Locke's person stages do not stand in the relevant transitive relation over time. A young man may remember an action he performed as a boy, the same man when old may remember performing an action as a young man and yet not remember the boy's action. It would seem, therefore, that the same man is not the same person.[11]

The view of the nature of a person that we have discussed so far is Locke's mature theory which was added to the second edition of the *Essay* of 1694. The first

---

[10] As Locke himself pointed out. See *E* II. xxvii. 20.

[11] Reid 2002, 276.

edition of the *Essay* includes a broader, composite notion of a *moral man*, that is 'a corporeal rational Creature' (*E* III. xi. 16). It is corporeal in so far as it has a body, but it is also rational. For Locke, it would not matter whether a moral man had the body of a monkey or some other creature. What counts is that it has reason and language that are adequate enough to be subject to law. In this case a rational monkey would be as much a moral man as a rational human: the shape and size of the corporeal body is an entirely contingent matter:

> were there a Monkey, or any other Creature to be found, that had the use of Reason, to such a degree, as to be able to understand general Signs, and to deduce Consequences about general *Ideas*, he would no doubt be subject to Law, and, in that sense, be a *Man*. (*E* III. xi. 16)

Yet it is a contingent fact that moral men have *human* bodies, bodies that are able to labour, to fight and to be imprisoned and even enslaved. And this brings us to Locke's conception of having a property in one's person.

I have stressed above that Locke's concept of person in the second and later editions of the *Essay* is independent of corporeal body: it is defined by a criterion of psychological continuity and is a forensic or moral notion. Yet, when Locke added this to the *Essay* he did not remove the short passage referring to the broader notion of the moral man, the corporeal rational creature. This broader conception is, of course, not incompatible with the later notion of a person and it is, perhaps, best for us to speak of a loose versus a strict notion of moral person in Locke's mature thought.[12] More importantly, however, the looser notion is the relevant notion for understanding a crucial claim in Locke's *Two Treatises of Government* which was published at the same time as the first edition of the *Essay*. This is the claim that 'every Man has a *Property* in his own *Person*' (*Two Treatises*, II. 27; Locke 1988, 287). As Kiyoshi Shimokawa (2006) has shown, to have a property in something, according to Locke, is to have an exclusive right of disposal. Moreover, for Locke, to have a property in one's own person is to have an exclusive right of disposal over one's life, health, labour and actions. If these are the objects of one's ownership, if this is what it is to have a property in one's person, then Locke's notion of a person in the *Two Treatises* is really equivalent to the broader notion of the *moral man*, rather than the narrower meaning of 'person' in *Essay* II. xxvii. And this has important implications for one's own self-diagnosis. We have an exclusive right of disposal of ourselves as corporeal rational beings and, thus, the responsibility for responding to our present and future states rests with ourselves.[13]

[12] See Mattern 1980.

[13] Locke never revised his notion of what it is to have a property in one's person in the light of his concept of person introduced in the second edition of the *Essay*. Nor does he seem to have made any attempt to incorporate his concept of property into the *Essay*. See Shimokawa 2006, 180.

8.3 The Self and the Understanding

We have examined Locke's distinctions between the soul, the understanding and the self. And we have also examined, albeit briefly, his different principles of individuation for masses of matter, living things and persons. It is natural to ask, therefore, what is the relation between the self and the understanding?

For Locke a person in the strict sense is 'a thinking intelligent Being, that has reason and reflection'. Later in Book Four he speaks of 'the *Idea* of our selves, as understanding, rational Beings' (*E* IV. iii. 18).[14] It is clear then that for Locke, to be a person is to use the understanding. We are not simply collections of past memories and static current conscious states. We are dynamic rational agents using reason and reflection with understanding. If this is the case, and it is hard to deny, then the forensic nature of the self is not exhausted by the evaluation of past and present actions. For, it will also involve fulfilling a set of epistemic duties in the way that one conducts their understanding. Some of these duties will have to do with the apportioning of evidence and the assessing of probabilities. Locke has strong views on the regulation of belief that are expressed in deontological terms.[15]

However, my concern is with another facet of Locke's account of the operation of the understanding, one that pertains, among other things, to natural philosophy, namely, the acquisition of real knowledge. Here is how Locke puts it:

> It is in the perception of the habitudes and respects our ideas have one to another that real knowledge consists; and when a man once perceives how far they agree or disagree one with another, he will be able to judge of what other people say, and will not need to be led by the arguments of others, which are many of them nothing but plausible sophistry. This will teach him to state the question right, and see whereon it turns; and thus he will stand upon his own legs, and know by his own understanding. Whereas by collecting and learning arguments by heart, he will be but a retainer to others; and when any one questions the foundations they are built upon, he will be at a nonplus, and be fain to give up his implicit knowledge. (*Locke Works*, 3: 236–237)

Locke brings this out clearly in his discussions of the value of reading for the conduct of the understanding. For example, in *Of the Conduct of the Understanding* he diagnoses the problem with voracious readers who 'make no great advances in real knowledge':

> The mistake here is, that it is usually supposed that by reading, the author's knowledge is transfused into the reader's understanding; and so it is, but not by bare reading, but by reading and understanding what he writ. (*Locke Works*, 3: 250, underlining added)

So, for Locke, one can acquire knowledge, but *real* knowledge is accompanied by understanding, or what I call intellectual grasp,[16] and it is the function of the

[14] Actually, in the first four editions of the *Essay* Locke had written 'rational Creatures'. The fact that he changed this to 'rational Beings' in the fifth (posthumous) edition is most likely an attempt to conform this passage more closely to the chapter on personal identity, *E* II. xxvii.

[15] See Wolterstorff 1996 for a thorough discussion. See also Corneanu 2011, 154–160.

[16] See Anstey 2013a, 321–325.

understanding to provide this real knowledge. It is natural, therefore, to ask how the understanding goes about acquiring such real knowledge.[17] Locke's answer to this question has both a diagnostic and heuristic side. The diagnostic side has to do with self-knowledge. It involves coming to terms with the weaknesses of our understanding.

> It will be no hindrance at all to our studies if we sometimes study ourselves, i.e. our own abilities and defects. There are peculiar endowments and natural fitnesses, as well as defects and weaknesses, almost in every man's mind. When we have considered and made ourselves acquainted with them, we shall not only be the better enabled to find out remedies for the infirmities, but we shall know the better how to turn ourselves to those things which we are best fitted to deal with. ('Of study', Locke 1968, 421)

As Corneanu has pointed out, Locke's *Conduct* provides an inventory of these defects. They include a lack of determinate ideas, seldom using one's own reason but relying on the reasoning of others, putting passion before reason, the association of ideas and considering only one side of a matter, that is, having a restricted vision (*Locke Works*, 3: 207–208, 275–278: cf. *E* II. xxxiii). Locke also claims that we lack sagacity in seeking out intermediate ideas. This supposed weakness is predicated on Locke's theory of demonstration as developed in the *Essay* and moves us from defects to remedies.

The *Conduct* also provides a critique of some inefficacious remedies. One of these is the Aristotelian syllogistic about which Locke claims: '[its] rules, that have served the learned world these two or three thousand years, and which, without any complaint of defects, the learned have rested in, are not sufficient to guide the understanding' (*Locke Works*, 3: 206). Once again, this claim is closely related to Locke's own theory of demonstration which he considers, in contrast to the syllogistic, to be a sufficient guide to the understanding.

So much for the diagnostic side of self-knowledge. Locke also provides, in the *Conduct* and elsewhere, a heuristic for self-knowledge and for the acquisition of real knowledge. His most succinct statement of the mechanics of coming to real knowledge is found in *Some Thoughts concerning Reading and Studying for a Gentleman*:

> the next step towards the improvement of his understanding must be, to observe the connexion of these ideas in the propositions which those books hold forth, and pretend to teach as truths; which till a man can judge whether they be truths or no, his understanding is but little improved; and he doth but think and talk after the books that he hath read, without having any knowledge thereby. And thus men of much reading are greatly learned, but may be little knowing.
>
> The third and last step, therefore, in improving the understanding, is to find out upon what foundation any proposition advanced bottoms; and to observe the connexion of the intermediate ideas, by which it is joined to that foundation upon which it is erected, or that principle from which it is derived. This, in short, is right reasoning; and by this way alone true knowledge is to be got by reading and studying. (*Locke Works*, 3: 294)

[17] Locke uses the term 'real knowledge' in a more constrained and technical sense in *E* IV. iv.

Note here the contrast between being learned and being knowledgeable. How is it that we become knowledgeable? First, by observing the connection between ideas that are constituents of propositions and to search for the foundations that these propositions 'bottom' upon. This is the process of right reasoning, that is, the operation of the understanding, and 'by this way alone true knowledge is to be got'. But Locke is at pains to emphasize one more point: right reasoning is performed and real knowledge is acquired when the understanding focuses on the way things are in themselves. As he puts it: 'To think of every thing just as it is in itself is the proper business of the understanding, though it be not that which men always employ it to' (*Locke Works*, 3: 235). Again, he tells us where clear and distinct ideas and words are to be found: 'he that will conduct his understanding right must not look for in the acuteness of invention, nor the authority of writers, but will find [them] only in the consideration of things themselves' (*Locke Works*, 3: 263). What Locke is prescribing here is, first, his theory of demonstration, second, the analysis of bottoming, and third, the observation of things themselves.

I have discussed Locke's theory of demonstration and his account of bottoming elsewhere and these need not detain us here (Anstey 2011, chapter 7). However, what is worth exploring is this notion of 'the consideration of things themselves'. For it is at this point that we see the deep connection in Locke's mind between the correct conduct of the understanding and the apprehension of nature. The whole thrust of Locke's view of natural philosophy—his rejection of speculative hypotheses, his emphasis on natural history and on observation and experiment, even his experimental history of the understanding itself—is an expression of the correct conduct of the understanding. Lockean self-diagnosis and Lockean natural philosophy are really two sides of the one coin.

8.4 Self-Diagnosis and Other Selves

For all of this self-diagnosis, for all the correcting of the diseases of the mind and the following of the way of ideas, for all the appropriation of past actions and of current consciousness, there is something missing from this account of Locke's view of the acquisition of knowledge of self and the world. For central to Locke's intellectual persona is the need for others and, in particular, the need for friendship.[18] And on this point one can hardly put it better than Locke does in a letter to William Molyneux of 26 April 1695:

> Meditating by ones self is like digging in the mine; it often, perhaps, brings up maiden earth, which never came near the light before; but whether it contain any mettle in it, is never so well tryed as in conversation with a knowing judicious friend, who carries about him the true touch-stone, which is love of truth in a clear-thinking head. (*Corr.*, 5: 351)

[18] See Yeo 2009.

Locke was highly regarded for his 'clear-thinking head' and this heightened the concern amongst his friends for his bodily wellbeing. William's brother Thomas Molyneux had earlier written to Locke expressing the widely held view that those of greater intellectual gifts seemed to suffer more in the body:

> I'm much concernd to hear You have Your Health no better, and on this occasion cannot but deplore the great Losses the Intellectual World in all ages has sufferd, by the strongest and sound Minds possessing the most infirme and sickly Bodys: certainly there must be some very powerfull Cause for this in Nature, or else we could not have so many Instances, where the Knife cutts the Sheath as the French materialy express it; and if so this must be reckond among the many other inseparable Miserys that attend Humane Affairs.
>
> I could wish the Physitiens Arte were so powerfull and perfect as in some measure to prevent so great an Evill. (20 December 1692, *Corr.*, 4: 598)

Locke replied with typical humility, thankful for Thomas' concern:

> I must acknowledge the care you take of my health in a way wherein you so kindly apply to my mind; and if I could persuade my self that my weak constitution was owing to that strength of mind you ascribe to me, or accompany'd with it, I should find therein, if not a remedy, yet a great relief against the infirmities of my body. (20 January 1692/3, *Corr.*, 4: 628)

Of course, this concern for the relation between body and mind was reciprocated on Locke's part and provides a reference point through which he related to his patients as a physician–philosopher. Mrs. Isabella Duke wrote to him on 21 January 1688:

> but 'tis true that the number of my infirmities increase, both of body, and mind; you only are capable of being physitian to both, come and try your skill; your presence will certainly cure the greatest disease of my mind, which languishes for want of it, your Art may be as successfull on the body too. (*Corr.*, 3: 339)

Sadly Locke's reply is no longer extant, though from Isabella's next letter it is clear that he spoke to both her mind and her body, for she speaks of 'your advice against my reading and tampering with physick … thus you call my faults to remembrance; as you do my infirmities of Body' (*Corr.*, 3: 382).

8.5 Conclusion

The theme of the interrelation between mind and body was nothing new in the seventeenth century: it was both a classical and a biblical motif. Nor was the cure of the mind an early modern innovation. Nevertheless, the emergence of experimental philosophy and the nexus of the study of natural philosophy and physic did accentuate the opportunities for and prevalence of informed self-diagnosis. Locke, being both a philosopher of the understanding and a practising physician, was probably England's most adept exponent of the holistic art of self-diagnosis.[19]

[19] I should like to thank Sorana Corneanu, Gideon Manning, Kiyoshi Shimokawa and the participants at the 'Testimonies: States of Mind and States of the Body in the Early Modern Period' at Caltech in May 2014 for comments on an earlier version of this paper.

Bibliography

Anstey, Peter R. 2011. *John Locke and natural philosophy*. Oxford: Oxford University Press.

———. 2013a. Locke on human understanding. In *The Oxford handbook of British philosophy in the seventeenth century*, ed. Peter R. Anstey, 310–328. Oxford: Oxford University Press.

———. 2013b. Review of Antonia LoLordo's *Locke's Moral Man*. *Mind* 122: 1146–1149.

———. 2015. John Locke and the philosophy of mind. *Journal of the History of Philosophy* 53: 221–244.

Corneanu, Sorana. 2011. *Regimens of the mind: Boyle, Locke, and the cultura animi tradition*. Chicago: University of Chicago Press.

Coventry, Angela, and Uriah Kriegel. 2008. Locke on consciousness. *History of Philosophy Quarterly* 25: 221–242.

Dewhurst, Kenneth. 1963. *John Locke 1632–1704: Physician and philosopher*. London: Wellcome.

Locke, John. 1823. *The works of John Locke*, 10 vols. London: Tegg.

———. 1968. In *The educational writings of John Locke*, ed. James L. Axtell. Cambridge: Cambridge University Press.

———. 1975. In *An essay concerning human understanding*, ed. Peter H. Nidditch. Oxford: Clarendon Press.

———. 1976–1989. *The correspondence of John Locke*, 8 vols, ed. Esmond S. de Beer. Oxford: Clarendon Press.

———. 1988. In *Two treatises of government*, ed. Peter Laslett. Cambridge: Cambridge University Press.

———. 1989. In *Some thoughts concerning education*, ed. John W. Yolton and Jean S. Yolton. Oxford: Clarendon Press.

LoLordo, Antonia. 2012. *Locke's moral man*. Oxford: Oxford University Press.

Mattern, Ruth. 1980. Moral science and the concept of persons in Locke. *Philosophical Review* 89: 24–45.

Reid, Thomas. 2002. In *Essays on the intellectual powers of man*, ed. Derek Brookes. Edinburgh: Edinburgh University Press.

Rickless, Samuel. 2015. Are Locke's persons modes or substances? In *Locke and Leibniz on substance*, ed. Paul Lodge and Tom Stoneham, 110–127. New York: Routledge.

Shimokawa, Kiyoshi. 2006. Locke's concept of property. In *John Locke: Critical assessments of leading philosophers*, 4 vols, Vol. 1, ed. Peter R. Anstey, 177–216. London: Routledge.

Stuart, Matthew. 2013. *Locke's metaphysics*. Oxford: Clarendon Press.

Thiel, Udo. 2011. *The early modern subject: Self-consciousness and personal identity from descartes to Hume*. Oxford: Oxford University Press.

Weinberg, Shelley. 2016. *Consciousness in Locke*. Oxford: Oxford University Press.

Wolterstorff, Nicholas. 1996. *John Locke and the ethics of belief*. Cambridge: Cambridge University Press.

Yeo, Richard. 2009. John Locke on conversation with friends and strangers. *Parergon* 26: 11–37.

Chapter 9
Joseph Glanvill on Imagination, Method and the Art of Thinking

Sorana Corneanu

Abstract In *The Vanity of Dogmatizing* (1661) and its subsequent versions (1665, 1676), Joseph Glanvill (1636–1680) sketched elements of an art of thinking in which the functions of the imagination played a central role. The root of his reflections was a defense of the immortality of the soul, carried out via an attack on faulty modes of thinking, while the end-result was an extended analysis of erroneous thinking combined with an Academic skeptical praise of free inquiry. My aim here is to reconstruct this line of Glanvill's thought, with an eye to its sources, its disciplinary rearrangements and its tensions, and to defend the following points: that the art of thinking in question here takes the form of a regulation of defective mental operations and is thus, contrary to a traditional logic, subsequent to, and dependent on, an analysis of error; that this analysis of error is a hybrid development, in which logical categories become mixed with psychological and medical notions; and that it is within the framework of his art of thinking that Glanvill develops his position relative to the correct method of inquiry in natural philosophy.

Keywords Joseph Glanvill · Pierre Gassendi · Kenelm Digby · Thomas White · Henry More · Francis Bacon · Pierre Charron · René Descartes · Immortality of the soul · Error · Imagination · Method · Art of thinking · Medicine · Logic · Skepticism

I would like to thank Gideon Manning for his feedback, encouragement and patience. I also acknowledge the support provided by the Honorary Fellowship of the Institute for Advanced Studies in the Humanities at the University of Queensland.

S. Corneanu (✉)
University of Bucharest, Bucharest, Romania
e-mail: sorana.corneanu@lls.unibuc.ro

G. Manning (ed.), *Testimonies: States of Mind and States of the Body in the Early Modern Period*, Archimedes 57,
https://doi.org/10.1007/978-3-030-39375-5_9

9.1 Introduction

As Joseph Glanvill explains in the "Epistle Dedicatory" that opens *The Vanity of Dogmatizing: or, Confidence in Opinions, Manifested in a Discourse of the Shortness and Uncertainty of our Knowledge, and its Causes* (1661), this work began life as a preface to a planned treatise on the immortality of the soul, which was never completed, since Henry More's "maturer undertakings" made it redundant.[1] The preface, however, initially drafted as "a Corrective of *Enthusiasm*, in a Vindication of the *use* of *Reason* in matters of Religion,"[2] became a thing of its own and was printed as such. The *Vanity* was quickly noted for its praise of various luminaries of the new philosophy, such as Francis Bacon, Pierre Gassendi, René Descartes, Kenelm Digby, Henry More and Thomas Browne; but its attack on Aristotelianism and defense of a skeptical attitude to dogmatic science provoked an angry response from the Roman Catholic Thomas White, entitled *Sciri, sive sceptices et scepticorum a iure disputationis exclusio* (1663), translated as *An Exclusion of Scepticks from all Title of Dispute* in 1665. White was himself an adept of new modes of philosophizing, and he embraced the variety that sought to integrate mechanism with Aristotelianism, in the manner of Kenelm Digby. The target of his attack was not only the *Vanity*'s anti-Aristotelianism, but also the "destructive contagion of Pyrronism"[3] he thought he detected in Glanvill's (as in Gassendi's) work. Glanvill's response, *Scire/i tuum nihil est: or, The Author's Defence of the Vanity of Dogmatizing*, was printed in 1665 together with the second version of the *Vanity*, re-branded as *Scepsis Scientifica: or, Confest Ignorance, the Way to Science*. The substance of the reply was to clarify the type of skepticism Glanvill purported to embrace and the uses to which it was put. The book was now dedicated to the Royal Society, of which he became a Fellow shortly after the imprimatur for the book was given in October 1664. A third version of the text appeared as the essay "Against Confidence in Philosophy, and Matters of Speculation," in Glanvill's 1676 collection of *Essays on Several Important Subjects in Philosophy and Religion.* Another essay in the same collection, "Of Scepticism and Certainty," rehearses and continues his controversy with White.[4]

The *Vanity* sequels were therefore, to a large extent, moves in a controversy over the issue of dogmatism versus skepticism. Yet, as we will see, while skepticism is an important context for understanding the core project of the *Vanity* texts, it is not the only one. I will propose that this project consists in drafting an art of thinking

[1] Glanvill 1661, A4r. Henry More's *Immortality of the Soul* was published in 1659.

[2] Glanvill 1661, A3v.

[3] White 1665, A2r.

[4] Medcalf 1970. On the Glanvill-White controversy, see also Cope 1956, 121–136; Spiller 1980, 16–23; Talmor 1981, 77–83; Lupoli 1984; Levitin 2015, 299–303, 371–372. White was Kenelm Digby's Catholic mentor and he wrote a continuation of Digby's *Two Treatises* (1644), which appeared in Latin in 1646 as *Institutionum peripateticarum ad mentem... Kenelmi equitis Digbaei pars theorica* and in English in 1656 as *Peripatetical Institutions in the Way of Sir Kenelm Digby*; see MacDonald 2013, 9.

well and avoiding epistemic error, out of which a set of recommendations for the correct method of natural inquiry also emerges. Skepticism did constitute a resource for this project, but so did the logical, psychological and medical literature of the time. Still, we should not forget that the original intention of the text was to pave the way towards a defense of the immortality of the soul. Even if this is by no means the prime concern of the *Vanity* texts as we have them now, echoes of the topic are preserved. It will be useful to review them in order to understand how it was that a book about the immortality of the soul became a book about thinking well, as well as why the imagination came to occupy such a prominent place therein. This will be the concern of the first section below. The second section will aim to disentangle Glanvill's account of error, its sources, disciplinary eclecticism and tensions. And the third section will comment on his positive art of thinking, the way it grows out of the analysis of error, and the way it leads to Glanvill's position on the method of natural inquiry.

9.2 Gross Apprehenders and the Immortality of the Soul

One aim of the *Vanity* is to argue for the illegitimacy of absolute (or, dogmatic) certainty in matters of human knowledge, owing to the natural limits of human minds and the faulty ways of our search for it—in other words, owing to human ignorance and error. Glanvill goes through instances of ignorance and error (chapters III–VI), the causes of ignorance and error (chapters VII–XV), an assessment of peripatetic philosophy (chapters XVI–XIX), an evaluation of the possibility of dogmatic science (chapters XX–XXII), a concluding indictment of dogmatism (chapter XXIII) and a concluding apology for (the new) philosophy (chapter XXIV). The first instance of ignorance and error he lists is the soul question: he considers the nature and origin of the soul, its union and relation with the body, and its faculties. On all of these issues, Glanvill claims, our knowledge is short and confused. The main strategy he follows in order to establish this is to engage in short surveys of the history of philosophy, whether ancient or modern or both, and to emphasize either the diversity of philosophical opinions (and hence the lack of consensus) or the explanatory shortcomings of otherwise ingenious theories.

Yet, there is one positive thesis that Glanvill obviously embraces and which runs through his discussion of the soul: its immateriality and immortality. There is little positive argumentation for it, however; presumably, this would have been reserved for the body of the planned treatise. What we have is primarily the castigation of those who deny the soul's immateriality for their short sight: what they do is form images of this object of knowledge in their imagination rather than allowing the intellect to cognize it, and thus their representations are tainted by the materiality of that faculty. As the *Scepsis* puts it, "possibly most have been deceived in this *Speculation*, by seeking to grasp the *Soul* in their *Imagination*; to which gross faculty, that purer essence is unpalpable: and we might as well expect to *taste* the

Sunbeams."[5] The question of the soul's action on the body, Glanvill continues in a further section, is easily solved if the soul is considered material, but those who opt for this solution are only "gross apprehenders," "conceiving [the soul] under some phancied appearance, and pinning on it materiall affections."[6] Instead, the immaterial soul's action on the material body should be recognized as the abstruse matter that it is, "of as difficil apprehension, as any mystery in nature".[7] The same goes for sensation, which raises the question of how the material body acts on the soul: the soul "is a stranger to such gross *substantiality* [as that of quantity and motion], and ownes nothing of these, but that it is cloathed with by our deceived phancies."[8] In the *Vanity*, this delusion of the imagination is attributed to the general disarray of the human faculties, among which a "shortness of our intellectual sight," following the Biblical Fall of Adam, to which Glanvill devotes the first two chapters.[9]

Glanvill also tackles this issue in *Scire/i*, his response to White's attack. White had countered Glanvill's points about the mystery of the soul's interaction with the body by offering the positive anti-dualist thesis that there is "no knowledge without a beat of phansies" and that the soul is not a distinct entity, but "the form or affection of the thing, Man."[10] Glanvill's reply to the latter point takes the form of yet another historical list of philosophical opinions, this time meant to show that the soul's distinction from the body and its immortality had been the belief of "all the wiser ages."[11] As to the former point, and against the "*Hobbian Hypothesis*," which he accuses White of upholding in his attack, Glanvill positively asserts that the soul is capable of knowledge "without an *impulse* from the *Phantasms*," in the case of its abstract speculations about God and other immaterial beings, on condition that "man puts off all the cloathing of the imagination." This is indeed "one of our chiefest Arguments for our *Immortality*."[12]

---

[5] Glanvill 1665a, 14; 1661, 18; 1676, 2.

[6] Glanvill 1665a, 16–17; 1661, 21.

[7] Glanvill 1665a, 16; 1661, 21.

[8] Glanvill 1665a, 21; 1661, 29.

[9] Glanvill 1661, 13. On the importance of the Fall of Adam to early modern views of natural philosophical knowledge, see Harrison 2007 (for Glanvill on this issue, see pp. 201–202). These introductory chapters are much shortened in the *Scepsis* and disappear altogether from the *Essays*.

[10] White 1665, 34, 35.

[11] Glanvill 1665b, 16. The list here includes the Chaldean oracles, Trismegistus, the ancient Jews, Pythagoras and Plato; the Egyptians, Chaldeans, Assyrians, Indians, Jews, the council of Rome; but also White himself, in the *Peripatetical Institutions*, and his sources, Digby and Aristotle (18–19). Compare with the lists in *Scepsis* and *Essays*, the purpose of which is to argue for the lack of consensus on this matter, hence for our ignorance and error, owing to the snares of our imagination: Plato, Aristotle, Hesiod, Anaximander, Heraclitus, Zeno, Xenocrates, the Egyptians, Chaldeans, Parmenides, Empedocles, Galen, Hippocrates, Varro, Thales, Crates, Decaearchus (the latter two only appear in *Scepsis*): 1665a, 13–14; 1676, 2. On the uses of the history of philosophy by proponents of the early modern new philosophies, see Levitin 2015 (as far as Glanvill is concerned, Levitin focuses on the relevance of this for Glanvill's views on the experimental method, 299–307).

[12] Glanvill 1665b, 29.

Glanvill is walking mined territory here. If by knowledge without an impulse from phantasms he means the intellect's direct access to immaterial objects, he upholds something he explicitly denies repeatedly in all the *Vanity* versions, where he states that immaterial beings are known by the intellect via their effects and attributes[13] and that there is indeed no human knowledge that does not proceed from the senses and thus that does not use material images.[14] Yet, "without an impulse from the phantasms" might refer to the intellect initially using the phantasms but leaving them behind in the process of knowing immaterial entities – a Scholastic tenet about the intellect's operations in the form it was repackaged by Gassendi, whose work Glanvill knew and admired.[15] In any event, Glanvill does not spell out his position on the operations of the intellect and their relation to the external and the internal senses. What he repeatedly points out instead is the imagination's negative role in blocking the intellect's grasp of immaterials.

The invective against "gross apprehenders," blinded by their material imagination, makes repeated appearance again in a tract on *The True Notion of a Spirit*, appended anonymously to Glanvill's *Saducismus Triumphatus*, starting with its second edition (1682; first edition 1681); we know now that the author of this tract was Henry More. More's spirit is both extended (against the "nullibilists") and immaterial (against the "Hobbists"). The extension in question is different from corporeal extension and must be understood in a different way. Therefore the materialists' failure to grasp this metaphysical extension "is to be imputed to their gross Imagination."[16] Glanvill's *Saducismus* is indeed a tract aimed at countering atheist materialism by proving, both empirically and rationally, that the immaterial realm exists, as manifested in the phenomena of apparitions, ghosts and witchcraft. It is thus the full expression of an intention which made only scattered appearance in the *Vanity* versions (whose purpose shifted along the way, as noted above).

The *Saducismus* also spells out something which the *Vanity* texts do not explicitly state, namely, the epistemological status of the tenet about the immortality of the soul. In keeping with the anti-dogmatic position which Glanvill made his profession of faith (on which more below), the proofs of the *Saducismus* are taken to indicate the high probability, not absolute certainty, of the existence of immaterial spirits.[17] It is plausible therefore that Glanvill offered the immortality of the soul as the most probable "hypothesis," and thus avoided the inconsistency of embracing an absolute positive truth while arguing for the pervasiveness of human ignorance and confusion.[18]

---

13 Glanvill 1661, 18; 1665a, 14; 1676, 2.

14 Glanvill 1661, 23, 67; 1665a, 18, 50; 1676, 16.

15 According to Gassendi, the intellect elevates itself to the knowledge of immaterial things by using the material images like a gradual stairway: Gassendi 1658, II, 441a, 447a; Bernier 1684, V.4, 282, 330–331. On scholastic views of the knowledge of the intellect, see Kessler 2007 (1988). On Gassendi, see Garber 1998, 771–772.

16 Anonymous [More] 1682, 134; v. also 137–140, 145, 154, 160.

17 Glanvill 1682, "Preface," n.p.

18 By "hypothesis" Glanvill means a "theory" that is held as probable and provisional (see third section below). In "Scepticism and Certainty" he distinguishes between an "infallible certainty,"

There is one other positive thesis put forward by the *Vanity* texts and it has to do with the definition of the soul in terms of its essential attributes, with the aim of showing that they do not share any common ground with the essential attributes of matter. Here Glanvill's debt to More is visible especially in the *Essays* version, where he lists "penetrable, indiscerpible and self-motive" as the essential attributes of the soul. But both *Vanity* and *Scepsis* used another list, which included "apprehension, judgment and discourse" as either "results" or "essentials" of the soul.[19] Now, apprehension, judgment and discourse were the names of what scholastic logic defined as the three acts of the intellect. The idea that they may be taken to define the nature of the soul could have been suggested to Glanvill by consulting Kenelm Digby's treatise on the soul in *Two Treatises* (1644), which is organized according to these three acts rather than to the traditional list of the faculties of the soul, and which, in addition, bases its defense of the immortality of the soul on an analysis of these three operations.

We will return to the three acts in the next section. For now, let us retain 'apprehension' as an important mental act. We have seen that the echoes of the concern with the immortality of the soul in the *Vanity* texts took the form of an indictment of the deceived apprehension of the imagination (which attributes material qualities to immaterials) and a call for the recognition of those things (such as the union and interaction of an immaterial soul with a material body) whose apprehension must remain difficult. As we will see in the third section, the positive counterpart to this will take the form of a regulation of apprehensions, coupled with a regulation of judgment. These, I will propose, are the core dimensions of Glanvill's art of thinking and the framework of his views on the correct method of inquiry in natural philosophy. Before we get there, however, we need to understand the shapes of Glanvill's account of error, which began with his interest in faulty thinking about the immortality of the soul, but came to incorporate much more, and more varied, material as it developed into a self-standing doctrine.

---

which we cannot attain, and an "indubitable certainty," which is our lot and consists in the firm (but not absolute) assent we grant to several things, such as the Cartesian cogito, our capacity to judge and reason, universal principles of thought, the evidence of sense and the testimony of others (1676, 47–49). The category of "indubitable certainty" seems to include here what Glanvill elsewhere calls "probability." (Other epistemic charts of the time are nevertheless more careful in distinguishing between the two. See Shapiro 1983). The immortality of the soul is not explicitly included here, but may well have been.

[19] Glanvill 1676, 4; 1661, 21; 1665a, 16. On More, see Garber 1998, 776–778.

9.3 The Causes of Our Errors: Logical, Medical and Skeptical Diagnoses

The condemnation of the imagination of "gross apprehenders" has an obvious platonic and neoplatonic flavor, and Glanvill's references to Plato and Plotinus in these sections duly register it.[20] But the range of Glanvill's sources is wider than this. Most interestingly, he uses the logical scheme of the acts of the intellect mentioned above, which he reshapes with the help of psychological and medical material; in addition, he draws on the skeptical analysis of cognitive error in its late Renaissance variant. As we will see, however, the issue of the mind's mistakes in knowing immaterial natures remains present here as well.

9.3.1 Logic Physicalized

The three acts of apprehension, judgment and discourse feature in Glanvill not only as essential attributes or modes of the immaterial soul, but also as the organizing scheme of his analysis of the errors owing to the imagination. In introducing his discussion in the *Vanity*, Glanvill writes that he shall "distribute *Intellectual operations* according to the known *triple* division, though with some difference of representation."[21] The *known* triple division of the acts or operations of the intellect was a commonplace of late scholastic logics. It was drawn from Aquinas's commentary on Aristotle's *On Interpretation* and was associated with, or sometimes replaced, the classic tripartite structure of terms, propositions and syllogisms. For instance, Robert Sanderson in *Logicae Artis Compendium* (1615) lists simple conception (of simple terms), composition and division (of propositions), and discourse or argumentation (to which he adds method as a related act). An alternative division can be found in Eustachius a Sancto Paulo's *Summa philosophiae quadripartita* (1609), which lists simple apprehension, judgment or enunciation, and reasoning or discourse.[22] The *difference* consists in the fact that the three operations – called by

[20] Glanvill 1661, 14–15, 70, 104; 1665a, 7, 53, 76; 1676, 16. Further on Glanvill's Platonism, see Cope 1956, ch. 6; Talmor 1981, 8–12; Lewis 2006.

[21] Glanvill 1661, 96; 1665a, 71. The extended analysis of the errors of the imagination in both *Vanity* and *Scepsis* is reduced to a single paragraph in *Essays*, where Glanvill now says that the topic is indistinguishable from that of the errors of the senses and (implicitly) of the understanding (1676, 22). Indeed the operations of these three faculties are closely intertwined in Glanvill's account of error, with the consequence that his official position on the distinction between intellect and imagination is seriously undermined. See below.

[22] Sanderson 1615, 2–3; Eustachius 1609, 20–21. Sanderson's was a popular logic textbook at Oxford, where Glanvill was a student in the mid- to late 1650s: see Feingold 1997, 294; Burns 2004, online edn., Jan 2008. http://www.oxforddnb.com.ezproxy.library.uq.edu.au/view/article/10790, accessed 6 April 2017. Eustachius was an acknowledged source for Gassendi: see Lolordo 2007, 35. Further on the logical literature on the three acts, see Ariew 2014, part II, chap. 1.

Glanvill simple apprehension, composition or judgment, and deduction or discourse or reason – are ascribed both to the intellect (or understanding) and to the imagination. As he puts it: "so we see, there is a triple operation of the *Phancy* as well as *Intellect*; and these powers are only *circumstantially* different."[23] The apprehension of the imagination is of absent corporeal objects (if the objects are present, the apprehension is performed by sense itself). The apprehension of the intellect is of "spirituals" (i.e., incorporeal entities) and occurs when "we denudate [the sensible ideas] of all material Phantasmes."[24] In judgment the mind puts together, or distinguishes between, apprehensions and thus forms propositions. If its objects are material, judgment is performed by the imagination; if immaterial, by the intellect. The connection of propositions and deduction of conclusions from them constitute the act of reasoning or discourse. Here Glanvill's distinction between imagination and intellect becomes evaluative: "when the conclusion is deduc'd from the unerring dictates of our faculties; we say the Inference is *Rational*: But when from misapprehended, or ill-compounded phantasmes; we ascribe it to the imagination."[25] All of this is odd, compared to the standard scholastic account, as well as to Digby; but it is not new, since Gassendi had already associated the three logical operations with the imagination. However, Glanvill's differences from Gassendi are equally notable.

In his *Physics* (in *Opera omnia*, 1658), Gassendi describes the three operations of simple apprehension, composition and division (also called assent and dissent, affirmation and negation, or enunciative judgment) and reasoning (also called argumentation, discourse, or inference) as functions of the imagination, which he takes to be the thinking faculty of the corporeal soul, as distinct from the intellect, the cognitive power of the rational soul. The ground of this unusual attribution seems to be the defense of *solertia brutorum*, the thesis that animals can think. Under this description, apprehension is understood as sensible perception of absent objects, judgment as a joining of the apprehended object with some non-abstract quality (e.g., a dog can form the judgment *man-master* by the mere collation of the object with the quality), and reasoning as a sequence of more than two apprehensions thus joined.[26] While there are suggestions both in the chapters on the imagination and in those on the intellect that the intellect also performs these three operations in its own distinct way, the three acts do not structure the discussion of the intellect (as they do that of the imagination). But Gassendi does use the three acts (plus 'ordering' as a fourth) as the organizing scheme in his *Logic*, where the distinction between intellect and imagination is all but erased.[27]

---

[23] Glanvill 1661, 98; 1665a, 72.

[24] Glanvill 1661, 97; 1665a, 71. This phrasing might indicate a Gassendist approach: see n. 15 above.

[25] Glanvill 1661, 98; 1665a, 72.

[26] Gassendi 1658, II, 411a, 413b; Bernier 1648, V.6, 229, 241. A digested summary is in Willis 1683, 38ff.

[27] See further Gassendi 1981, li–lvi.

According to the *Logic*, simple apprehension can also be called "simple imagination," since its object is the image the mind thinks about and as it were looks on; as such, "it is sometimes taken for the imagining faculty, which the Greeks call *phantasia*, and is attributed to the inferior part of the soul, which is common to men and brutes."[28] Moreover, simple apprehension can be an act of compounding images, for example in forming the idea of a *golden mountain* or of a *centaur*.[29] As far as the second act is concerned, Gassendi works with a double sphere of this act: on the one hand, it consists in a composition of two ideas that are mutually agreeable; on the other, it is considered under the aspect of predication, where something is affirmed or denied of another thing. Still, the resulting proposition is also called a 'complex imagination'.[30] The *Physics* makes explicit the continuity between the first two acts, and the porous line between intellect and imagination: the joined apprehensions of the second act (say, the dog's *man-master*) are not different from the composition of the imagination by which it obtains such simple apprehensions as the *centaur* or the *golden mountain*; in both cases the ground of these operations is the joining together of impressions in the brain.[31] Indeed, Gassendi adds, this holds true of the operations of the intellect as well: the second act (a propositional judgment) is of the same nature as the simple apprehension "carried out by composite terms," such as *rational animal* or *just man*.[32] Finally, the third act, of reasoning, is understood as a sequence of more than two apprehensions joined together, based on their agreement.[33] In this way, the third act seems no more than an extended second act, i.e., a more complex chain of judgments, which, in turn, is an extended chain of simple apprehensions.

Although Gassendi's treatment of the three acts swerves significantly from late scholastic logic, it is in fact convergent with the discussions of the imagination in late scholastic physics, in the sections devoted to the soul (the *de anima* sections). Indeed, the late scholastic physics make room for the three acts as functions *of the imagination*. The context is the discussion of the internal senses, i.e., those faculties which humans share with brutes; here we are to take *phantasia* as a unified internal sense, subordinating all the other sensitive faculties of the medieval tradition (the aestimative, the cogitative and memory). As such, it is also called "particular reason," which humans and animals have in common, although in humans it functions in conjunction with the intellect. And just like the intellect, the imagination is capable of all the three acts. The Coimbrans, for instance, state that *phantasia* is indistinguishable from the *cogitativa* and "can compose propositions out of singular terms and can discourse pertinently on their objects."[34] Eustachius writes in a similar

---

[28] Gassendi 1658, I, 92b; Bernier 1648, I.1, 6.

[29] Gassendi 1658, I, 93a; Bernier 1648, I.1, 12.

[30] Gassendi 1658, I, 99a; Bernier 1648, I.1, 53.

[31] Gassendi 1658, II, 410a; Bernier 1648, V.6, 225.

[32] Gassendi 1658, II, 411a; Bernier 1648,V.6, 231.

[33] Gassendi 1658, II, 413b; Bernier 1648, V.6, 241.

[34] Collegium Conimbricense 1600, 401.

way of the *aestimativa*, which he sees as a specification of *phantasia*[35]: it "composes, divides, makes propositions, and pertinently discourses in its own way about singulars."[36] It is also in the late scholastic physics that the division lines are blurred between the composition of apprehension and that of judgment, and between those specific to the imagination and those specific to the intellect.[37] Thus, Gassendi's departures from the scholastic logical approach to the three acts is in fact made possible by the approach to the same topic in late scholastic physics. The tensions in Gassendi are therefore a legacy of the tensions in the scholastic distribution of the topic of the three acts between their logic and their physics.

Glanvill uses a Gassendist approach but further modifies it, since his main interest is not in an account of the cognitive operations in their normal operation, but rather in an account of how they go wrong. In other words, his concern is not with the soul of brutes, but with the depraved souls of men. The assignation of error to the imagination is not alien to the late scholastic context: the malfunctions of the imagination, including on account of its relation to the body, are treated in the context of the *de anima* sections of a physics.[38] The theme is however much more developed in a Renaissance eclectic context, where the Aristotelian, stoic and skeptical legacies combine in such a way that the imagination becomes the faculty which assents to *misapprehended* things.[39] The early modern logics had already made room for such material. In *De augmentis scientiarum* (1623), Bacon had posited that errors are owing partly to "false forms of demonstration and ill-ordered methods of reasoning,"[40] which constitute the traditional doctrine of fallacies, and partly to the errors and weaknesses of the mind itself, which are the substance of his new doctrine of the idols of the mind. He described the latter as being rooted in the "corrupt and ill-ordered predisposition of the mind, which as it were perverts and infects all the anticipations of the intellect." Here both the imagination and the mind's union with the body become relevant, while the analytical doctrine of proof and demonstrations "is not to be looked for."[41] Gassendi paid attention to this topic, too (very likely under Bacon's influence), as he included several canons (XI to XIV) on the question of deception in the first part of his *Logic*, devoted to simple apprehension. He mentions – without elaboration – deceptions owing to the senses (even if, following Epicurus, he maintains that sense itself cannot be mistaken), as well as to temperament, disposition of mind, custom, preconceptions, authority, ambiguous words and figurative language.[42] Glanvill touches on all of these (except the latter

[35] Eustachius 1609, 394.

[36] Ibid., 404.

[37] I develop this topic in a forthcoming paper. A remarkable text from this perspective is Velcurio 1537.

[38] E.g., Ibid., 29.

[39] E.g., Pico 1501: see Park 1974, 146.

[40] Bacon, *De augmentis scientiarum,* in Bacon 1963–1994 (1857–1874), IV, 412.

[41] Ibid., 431. On Bacon's idols and the imagination, both epistemic and physiological, see Corneanu and Vermeir 2012.

[42] Gassendi 1658, I, 96a–97b; Bernier 1648, I.1, 32–40.

two) in much more elaborate detail under 'judgment' and 'reasoning', and makes the imagination central to the account.

In short, following Bacon and Gassendi, Glanvill's move is to reshape the scholastic logical format of the three acts by bringing in material traditionally treated under physics, both relative to the neutral description of the functions of the imagination and in terms of the errors to which it is prone. The move, I suggested, did not really do violence to the scholastic inheritance, but rather made manifest tensions already latent in it. But that material, as found in a scholastic physics, was not as rich as the elaborations found in the moral-psychological and medical tracts on the epistemic misbehavior of the human mind that proliferated in the early modern period. We will survey these debts below, after a review of Glanvill's account of the errors of the mind.

Glanvill devotes a substantial chapter to the errors of the senses, but closes it by absolving them of the criminal charge. He does so by means of the argument that the senses simply "give us the truth of their *Sentiments*" and the problem lies rather with the precipitance or rashness of the understanding's assent[43] (which, as we will see, is under the influence of the imagination). The purpose of this move is to shift the focus of attention in the analysis of error from the external to the *internal* senses: "and so though they [the senses] are some way accessory to our delusion; yet the more principal faculties are the *Capital offenders*."[44] By "principal faculties" Glanvill means the imagination and the understanding. The simple perception of the imagination is as innocent as that of the senses. But in its performance of judgment, the imagination is, he writes, "the almost fatal means of our deception, through the unwarrantable *compositions, divisions,* and *applications*, which it occasions the *second Act* to make of the *simple Images*."[45] The example he gives is of the enthusiast, who judges that the voices and visions in his head are external realities. The case is similar with that of the person who thinks that her perception of the quiescence of the earth depicts an external reality.[46] In both cases, it is not the simple perception itself (of the senses or of the imagination) that is at fault, but the judgment made by our minds of this simple perception. But there is one more ingredient in the second act: in the case of the enthusiast, the error of judgment is owing to the "strong Idea's of [the voices and visions], being conjur'd up into the Imagination by the heat of the *melancholized* brain.[47]

If we look closely at this explanation, it appears that there is in fact a double operation at play here. One is that the imagination hosts 'ideas' or 'apprehensions' that acquire an extra quality of 'strength' owing to the melancholic brain. The other is that the imagination performs the erroneous judgment that those ideas faithfully

[43] Glanvill 1661, 92–93; 1665a, 67–68. He is here of a mind with Gassendi, but his reference is to Augustine's *Contra academicos*, and thus implicitly to Cicero's *Academica* and his reports of stoic and skeptical approaches to knowledge.

[44] Glanvill 1661, 92; 1665a, 67.

[45] Glanvill 1661, 99; 1665a, 72.

[46] Glanvill 1661, 92–93; 1665a, 67–68.

[47] Glanvill 1661, 99; 1665a, 73.

represent external reality. This judgment is somehow prompted by the strength of the mental image. What is identified here is the mind's capacity of being impressed in such a way that the images it holds and works with come to dominate one's internal theatre so that they force assent and thus determine the formation of erroneous belief. Glanvill thinks that this capacity is best explained as a function (or malfunction) of the imagination, conjoined with the action of a distempered body on the mind.

He continues his list of examples with the "hypochondriacal imaginations," with men who think they are emperors or that they are made of glass – which are standard examples in early modern accounts of melancholy. But in the same category are also erroneous philosophico-theological tenets about immaterial beings, such as that angels occupy a place, or that God can be attributed human faculties. The inferences and deductions based on such judgments, so the third act (of deduction), can only be as false as they are. Falsity, however, is not the only danger: obstinacy and credulity compound it, and they are both explained as products of, respectively, "a tenacious *Imagination*"[48] and of "its impressing a strong perswasion of the Truth of an Opinion, where there is no evidence to support it."[49]

Thus, the trouble with a malfunctioning imagination is not only that it judges and infers wrongly, but that it becomes obstinate in its tenets. Such obstinacy – the strength of adherence to a belief – mirrors the vigor with which images come to be impressed on the imagination in the first place. The "strong ideas" of the first apprehension and the "strong persuasions" following the judgments and inferences based on them are effects of the work of the imagination. The general effect of these "vigorous impressions of *Phancy*"[50] is a sort of freezing of the mind, now occupied by these images and beliefs.

The chapter on the understanding is a variation on the account in the chapter on the imagination. Unlike Gassendi, Glanvill does not pursue the thought that the intellect has its own operations, divorced from matter. Instead, the actions of the intellect refer here indiscriminately both to sensible and to insensible objects, and the concern is with "the *forwardness* of our *Understandings assent*, to slightly examin'd *conclusions*, contracting many times a firm and obstinate belief from weak inducements; and that not only in such things, as immediately concern the sense, but in almost every thing that falls within the scope of our enquiry."[51] A similar idea was rendered a couple of pages earlier by means of talk of the strong persuasions of the imagination. The focus here, however, is in fact on the assent of the *understanding*, the forwardness of which Glanvill explains by invoking the body and the imagination: it is because we have "embodyed souls" that even the abstracted work of meditation cannot be performed "without a considerable proportion of

---

[48] Glanvill 1661, 103; 1665a, 75.

[49] Glanvill 1661, 104; 1665a, 76.

[50] Glanvill 1661, 105; 1665a, 77.

[51] Glanvill 1661, 106–7; 1665a, 78.

Spirits to assist in the Action."[52] Besides the "*volatile* nature" of our bodily spirits, which are hard to retain "in any long service, but do what we can, they'l get loose from the Minds *Regimen*," we have to contend with "the excursions of our roving *phancies* (which cannot be kept to a close attendance)."[53] Compared to the chapter on the imagination, where the focus was on the freezing of the mind in obstinacy and credulity, here at stake is the unruly, precipitous, roving nature of the imagination, which, together with the swift-moving bodily spirits, leads to the misplacement of the mind's assent, and thus to erroneous judgment.

If we put together the accounts in these two chapters, on the imagination and on the understanding, we obtain two noteworthy results. One is that there is rather confusing overlap between the judgment of the imagination and the judgment of the understanding, as they are involved in the operations of image-composition, predication and assent. The other is that the scope of 'apprehension' is larger than initially defined: it can be a perception of a thing (e.g., "voices") but it can also be a propositional conception (e.g., "that angels occupy a place"). In light of all of this, this is what the sequence of apprehensions and judgments would look like in the case of the "voices" in one's head: the simple apprehension of "voices" acquires strength from the imagination and the body; it will therefore be combined with the quality "real existence" by the imagination, which results in the collation "voices-real existence," but also takes the form of the proposition "voices are external realities" (a judgment seemingly attributed to the imagination); the propositional apprehension thus obtained also acquires strength from the imagination and the body; it will therefore induce assent, whether in the form of the persuasion of the imagination or of the approbation of the understanding, which is additionally intensified by the swift movement of the imagination and the body; the end-result is the strong, obstinate belief in the proposition that "voices are external realities." And this is how an enthusiast is made.

All of this is compounded by the intervention of the mind's passions, desires and pleasures. The theme is pursued in the three subsequent chapters devoted to the errors of the affections, where the Baconian-Gassendist topics of custom, education, interest, love of one's own opinions and misplaced reverence for others' authority or prestige are rehearsed as so many species of passion-induced errors. The same chapters introduce a further medico-psychological theme: when treating of the affection of self-love, Glanvill rehearses the popular Galenic tag that had known a long career since the sixteenth century, that "the *complexion* of the *mind*, as well as *manners*, follows the *Temperament* of the *Body*"[54] and associates it (as had often happened, including in Galen's own tract) with the idea of the diversity of wits: "And I believe the Understanding hath its *Idiosyncrasies*, as well as other faculties. [...] And the

[52] Glanvill 1661, 109; 1665a, 80.

[53] Glanvill 1661, 111; 1665a, 81.

[54] Glanvill 1661, 121; 1665a, 89. Galen's *Quod animi mores corporis temperamenta sequantur* made an important impact on late Renaissance and early modern medical and philosophical thought about the cure of minds, following its publication in Latin in the first half of the sixteenth century: see Savino 2011; Corneanu 2017.

opinions, which are suited to their respective *tempers*, will be sure to find their welcome, and to grow without manure."[55] Such diversity is best explained via the influence of the temperamental complexion on the imagination.[56] This of course raises problems for the thesis of the immortality of the soul, which Glanvill does not address.

In sum, Glanvill's analysis of error loses any connection with the formal sort of logical analysis and becomes an entirely psychologico-medical affair. For him, a neutral account of the three acts, even if looking at mental operations rather than terms, propositions and syllogisms, proved unequal to the task of rendering a full picture of what the mind can do. The strong, tenacious impression and the swift, precipitous assent were modalities of apprehension and judgment that the traditional theory could not account for, but which Glanvill thought were essential to understanding the errant behavior of the mind in natural inquiry. The right resources for talking about this were to be found not so much in logics (unless they were of the new kind, like Bacon's and Gassendi's), but in physics – that is to say, either in the *de anima* sections of late scholastic physics (where such things were present *in nuce*) or, in much more elaborate form, in the philosophical, psychological and medical tracts on the failures of the mind that were in vogue during his time. Three such failures stand out in Glanvill's text: enthusiasm, melancholy and dogmatism itself. In what follows I will look at what must have been Glanvill's main sources on each of these heads; I will also note their own engagement (whether explicit or not) with the three logical acts as well as their debt to medical thought.

9.3.2 *Enthusiasm, Melancholy and Dogmatism*

In the case of enthusiasm our task is easy, since Glanvill has an explicit reference in his text to Henry More, his model on the question of the immortality of the soul and his long-time collaborator on the topic of ghosts and apparitions. In this context, however, the reference must be to More's *Enthusiasmus Triumphatus* of 1656. Section VI of that work is entitled "The enormous strength of Imagination the Cause of Enthusiasme" and in it More talks about the way the soul "so farre sinks into Phantasmes that she cannot recover her self into the use of her more free faculties of *Reason and Understanding*" and thus is brought to "believe a lie" owing to the strength of the imagination. Here is how he analyses the cases of people who think they are God, the Messiah, the Holy Ghost or the angel Gabriel:

> For their conceptions are not so pure or immateriall, nor solid or rational, but that these words to them are alwaies accompanied with some strong Phantasme or full Imagination; the fulnesse and clearnesse whereof [...] does naturally bear down the Soul into a belief of the truth and existence of what she thus vigorously apprehends: and being so wholly and entirely immersed in this conceit, and so vehemently touched therewith, she has either not

[55] Glanvill 1661, 122; 1665a, 90.

[56] Glanvill 1661, 124; 1665a, 91.

> the patience to consider anything alledged against it, or if she do consider and find her self intangled, she will look upon it as a piece of humane sophistry [...].[57]

Here More diagnoses the failures of the operations of the soul by pointing to the same qualities of the three acts as in Glanvill: on the one hand, the strong image, vigorous apprehension and vehement impression; on the other, the impatient belief and the faulty reasoning proceeding from it. While in this passage the imagination does not appear explicitly, in the subsequent paragraphs it is singled out as the blameworthy faculty: More adds that the fortifying of absurd phantasms can also be the effect of custom and education, but that the imagination works much more swiftly to this end.[58] The next sections are devoted to the responsiveness of the imagination to diets and natural substances, and thus to its physiological aspects.[59]

We can compare this with another important work on the subject, published 1 year before More's, Meric Casaubon's *Treatise of Enthusiasme* (1655). Casaubon comments on the overbearing power a strong and violent imagination has on the understanding's assent:

> Neither is every error of the imagination an error of the understanding. [...] But if once the imagination be so strong & violent, as to force assent from the understanding, so that no power of ratiocination that is left in us, is strong enough to make us believe that it is otherwise than we imagine: is not this a depravation of the Understanding, as well as of the Imagination?[60]

The immediate context of this remark is a discussion of a controversial point stemming from Galen's description of phrenitis in *De locis affectis* and *De symptomatum diferentiis liber 4*: the question about which faculty was affected in various cases of phrenitis, whether the imagination, the understanding or both. The latter work discusses the case of one Theophilus (in whom the imagination was hurt but the judgment remained intact), which Casaubon rehearses in his text.[61] This adds another complication to the difficulties around the act of judgment noted above: the logical understanding and the Galenic understanding are rather different things, the latter much closer to the cogitative function of the sensible soul in the scholastic tradition (whether called the cogitative, imaginative or estimative faculty). We are thus brought into the territory of physics once again, this time via medicine. The same Galenic place is discussed by Gassendi in his *Physics* book on the imagination, where he draws the conclusion that, despite such examples, imagination and judgment (and memory) are not necessarily very different faculties,[62] and all the internal senses may well be reduced to the imagination alone, as the cognitive faculty of the

[57] More 1656, 4–5. See further Heyd 1995; Crignon-De Oliveira 2008, 323–335.

[58] More 1656, 5.

[59] See further Vermeir 2008.

[60] Casaubon 1655, 88–89.

[61] Casaubon 1655, 88. Galen, *De symptomatum diferentiis liber 4*, in Galen 1821–33, VII, 60–61; *De locis affectis* 4.2, in ibid., VIII, 225.

[62] Gassendi 1658, II, 403a; Bernier 1648, V.6, 196–197.

sensible soul.[63] Gassendi also comments on the quality of strength that characterizes the apprehensions of the imagination (as seen in passions), and analyzes it in physiological terms as a result of the strong movement of the bodily spirits.[64]

As far as melancholy is concerned, the obvious reference must be Robert Burton's *Anatomy of Melancholy* of 1621. Burton takes over from the very popular *Passions of the Minde in Generall* (1604) by the Jesuit Thomas Wright the passage that describes the mechanism through which, in passion, the imagination influences both the conception in the mind and the behavior of the corporeal spirits and humors, but adds several qualifications to underline precisely the intensity of the apprehensions of the imagination:

> To our imagination commeth, by the outward sense or memory, some object to be knowne (residing in the foremost part of the braine) *which he mis-conceaving or amplifying*, presently communicates to the Heart, the seat of all affections. [...] *If the Imagination be very apprehensive, intent and violent*, it sends great store of spirits to, or from the heart, *and makes a deeper impression, and great tumult,* as the humours in the Body be likewise prepared, and the temperature it selfe ill or well disposed, the passions are longer and stronger.[65]

This "force of imagination," as Burton calls it, is the main cause of melancholy. His explanation is partly physiological (it is moved by humors), partly psychological: it works by "keeping the species of objects so long, mistaking, amplifying them by continuall and strong meditation, until at length it produceth in some parties reall effects, causeth this and many other maladies."[66] To further this point, Burton lists melancholy together not only with other passions and affections, but also with dreams and ecstasies, as well as the imprint of the mother's imagination on the fetus and other such strange phenomena, from transmission of yawning to fascination. All these, he concludes, are explained by the "strong conceipt of the imagination," which is "*astrum hominis*, and the rudder of this our ship, which reason should steire, but overborne by phantasie, cannot manage, and so suffers it selfe and this whole vesell of ours to be overruled, and often overturned."[67]

Another work worth invoking here is the *Treatise of the Passions and Faculties of the Soule* (1640) by the Calvinist divine Edward Reynolds. The *Treatise* was famous at the time, and was dedicated to the same princess Elisabeth of Bohemia who also prompted Descartes's own *Passions*. Reynolds's main interest is in troubles of mind and vexations of spirit, but also in heresy and false prophecy, together with the vices of credulity and obstinacy. The corruptions of the imagination responsible for these are mainly of two kinds. One is its "heaviness and sluggish fixedness," which is a "peremptorie adhesion and too violent intension in the Fancie on some particular objects," and which results in a sort of dullness of mind, sometimes

---

[63] Gassendi 1658, II, 402a; Bernier 1648, V.6, 193.

[64] Gassendi 1658, II, 409b; Bernier 1648, V.6, p. 221.

[65] Burton 1989–1990, I, 249 (italics mine, to reflect Burton's additions to Wright's text); Wright 1604, 45.

[66] Burton 1989–1990, I, 250.

[67] Burton 1989–1990, I, 254.

intensified to the point of madness; an example is the "depravation of the Phantasie in melancholy men."[68] Another type of corruption is "the *Levitie* and too much *Volubilitie* of this Power, proceeding from the over-hastie obtrusion of the *species*," on account of which the imagination "makes many needlesse excursions upon impertinent things."[69] The phenomenon is explained as an excessive quickness of apprehension and mobility among objects of apprehension, resulting in both false representations and hasty reasoning upon such shallow grounds. The way Reynolds makes this point is interesting, since it signals the logical connection in a rather overt manner:

> [Quickness of apprehension,] though it may seeme to be the most peculiar worke of Reason, yet the Imagination hath indeed the greatest interest in it: For, though the Act of Apprehending be the proper work of the Understanding, yet the forme and qualitie of that Act (which properly makes it a Thought in that strict sense, wherein here I take it) namely, the lightnesse, volubilitie, and suddennesse thereof, proceeds from the immediate restlessnesse of the Imagination; as is plaine, by the continuall varietie of Dreames and other Fancies, wherein the Facultie is the principall worker.[70]

Ganvill would have agreed: while officially apprehension is an act of the intellect, as logic teaches, the explanation of the particular qualities of this act – its quick mobility, or else its fixation on an object – requires the metamorphic capacities of the imagination, of which (Reynolds does not say this, but it is one of the outcomes of our review here) it is physics and medicine that can inform us.

Finally, regarding dogmatism, one probable source for Glanvill's attack on the imagination from an anti-dogmatic vantage point was Pierre Charron. Glanvill explicitly cites Charron on the topic of free inquiry,[71] to which we will come back in the next section. It is however likely that Charron on the imagination was as influential, since the two topics are as closely related in Charron as they are in Glanvill.

"Of the imagination and opinion" is one of the chapters describing the "natural consideration of man" in Book I of Charron's *Of Wisdome Three Bookes* (1606; French original, *De la sagesse*, 1601). It consists of material culled from Montaigne's *Essais* (1580–1588) and is a nice digested summary of the typical magical, medical and epistemological aspects of the humanist renaissance view on the imagination. Of particular interest here is Charron's treatment of the neo-stoic and skeptical theme of the imagination as harbor of opinion. Here is the relevant passage:

> In this part and facultie of the soule doth opinion lodge, which is a vaine, light, crude and imperfect iudgement of things drawen from the outward senses, and common report, setling and holding it selfe to be good in the imagination, and neuer arriving to the understanding, there to be examined, sifted, and laboured; and to be made reason, which is a true, perfect and solide iudgement of things: and therefore it is uncertain, inconstant, fleeting, deceitfull, a very ill and dangerous guide, which makes head against reason, whereof it is a shadow

68 Reynolds 1640, 29.

69 Reynolds 1640, 28.

70 Reynolds 1640, 23–24. See further Gowland 2008, 103–117; 2012, 10–38.

71 Glanvill 1661, 32; 1665a, 172. This is removed from *Essays*, together with the whole discussion of the imagination. Maybe this double removal points to the connection between the two topics.

> and image, though vaine and untrue. It is the mother of all mischiefs, confusions, disorders: from it spring all passions, all troubles. It is the guide of fooles, sots, the vulgar sort, as reason of the wise and dexterious.[72]

'Opinion' (Lat. *opinio*; Gr. *dogma*) as the cause of passions, as well as the association of opinion and imagination are common topics in the Renaissance literature on moral and cognitive error that is indebted to both the stoic and the skeptical traditions.[73] As Katharine Park explains, the attribution of error to the imagination in this context is owing to a "persistent mistranslation of the relevant texts": the Greek *φαντασία* in the stoic and skeptical sources (mainly Epictetus and Sextus Empiricus), meaning the primary sense impression to which the mind gives its assent in forming a belief, is translated in the sixteenth century as *phantasia* or *imaginatio*, resulting in a confusion between the faculty and its object.[74] The faculty of the imagination becomes thus responsible for the formation of the erroneous opinion, and acquires the functions of judgment/assent. Moreover, the difference between its work and that of the understanding, or of reason, becomes evaluative rather than descriptive.

For Charron, 'opinion' thus understood is one of the maladies of man's mind (*esprit*), which accounts for the enslavement of minds to unexamined opinions taken as absolute truths, or principles, on which they found their thinking – the substance of dogmatism and credulity. The mind becomes "preoccupied" and "tainted" with "certain fantasicall opinions" and "ever afterwards followeth and iudges according to them, without regard either of farther enquiry, or recoiling backe."[75] This analysis is extended into the "moral consideration of man" in Book I, under the head of "presumption." Presumption with respect to man's opinions takes the form of three kinds of folly: the first includes credulity (a facility to believe) and incredulity (the rash rejection of things not easily understood); the second is obstinacy and it follows from the first; and the third, which proceeds from the other two, is dogmatism (the imposition of unexamined "principles" on others).[76] The counterpart to this diagnosis is Charron's defense of the freedom of judgment in Book II, which he counts as one of the dispositions to wisdom (*sagesse*). The Charronian freedom of judgment is what guarantees the "most happie estate of the spirit, which by this meanes keepeth it selfe firme, upright, constant, inflexible, alwaies free and to it selfe" and by which "a man is in the calme, under covere, and out of danger of participating so many errors produced by the fantasie of man."[77]

In what follows we will look at Glanvill's variant of the Charronian – which is to say, Ciceronian and, ultimately, Carneadean – freedom of judgment. We will also note how an Academic-skeptical approach to the formation of belief comes to be

---

[72] Charron 1606, 67–68.

[73] Park 1974, chapter 4.

[74] Ibid., 137–140.

[75] Charron 1606, 63.

[76] Ibid., 157–160.

[77] Ibid., 241.

grafted onto the logical scheme that guided Glanvill's account of error and how Glanvill's views on method in natural inquiry grow out of this graft.

9.4 Method and the Freedom of Inquiry

All the three versions of the *Vanity* conclude with a rehearsal of the main "considerations against dogmatism" developed along the way, now condensed under six heads.[78] The third and fourth touch on the social effects of dogmatism: it is a disturber of peace (since it fuels schisms and heresies, in which the obstinacy of 'opinions' is at its highest) and it breeds ill manners.[79] The first two and the latter two, instead, speak of the dogmatic attitude itself and the cast of mind it rests on. Here Glanvill openly relies on neostoic and skeptical themes. Dogmatism "dwells with untamed *passions*," hence it is the direct enemy of the (stoic) "Art of *Autocrasy*" and of the government of judgment, "the *Hegemonical* power."[80] Dogmatism also betrays ignorance, or else a poverty and narrowness of mind, which keeps it forever captive in its own circle of errors, and is thus the exact opposite of the modesty and caution in the pursuit of knowledge that characterize the "generous soul[s]," "nobler spirit[s]" or "*larger* Minds," whose "temper of Mind" makes them fit for "true Philosophy and right Knowledge."[81] The paragraph concluding this discussion in the *Vanity* makes Glanvill's debts here transparent:

> The determinations of the nobler spirit, are but temporary, and he holds them, but till better evidence repeal his former apprehensions. He won't defile his assent by prostituting it to every conjecture, or stuff his belief with the luggage of uncertainties. The modesty of his expression renders him *infallible*; and while he only saith he *Thinks so*, he cannot be deceiv'd, or ever assert a *falshood*. But the wise Monseur *Charron* hath fully discourst of this *Universal liberty*, and sav'd me the labour of inlarging. Upon the Review of my former considerations, I cannot quarrel with his *Motto*: in a sense *Je ne scay*, is a justifiable *Scepticism*, and not mis-becoming a Candidate of *wisdom*. *Socrates* in the judgement of the *Oracle* knew more then *All men*, who in his own knew the least of *any*.[82]

The freedom of judgment, we have seen, was one of Charron's "dispositions to wisdom." As he explains in the second chapter of Book II of his *Sagesse*, it consists in three things: "to iudge of all things," by which he means to examine and weigh reasons on every side of a question, rather than to affirm and determine; "not to be married or bound to any" affirmation or determination; and "to continue open and readie for all" judgments, that is, "alwaies to hold [oneself] readie to entertain better

[78] *Vanity* and *Scepsis* also have an "apology for philosophy" after these "considerations," which is an extension of the same points.

[79] Glanvill 1661, 228–232; 1665a, 168–171; 1676, 30–31.

[80] Glanvill 1661, 227; 1665a, 167; 1676, 30.

[81] Glanvill 1661, 225–227; 1665a, 165–167; 1676, 29–30, 33.

[82] Glanvill 1661, 234–235; 1665a, 171–172.

if it appeare."[83] These form the "universal" spirit of the wise man, or of the "modest, discreet, and temperate searcher of the truth and true Philosophie," against the "particular" mind of "weake people" and "obstinate opiniative affirmers."[84] In elaborating the second point, Charron explains that the search for truth which maintains the mind free is one that refuses the closure of certainty and only admits that some opinions are more probable (*verisimilia*) than others, that in certain cases "*ita videtur*, it seemeth so."[85] Since certainty of the absolute kind is not available to us, the willingness to embrace "certaine principles" is nothing more than an "uniust tyrannie" over our own minds and those of others, and "from thence do spring pride, ambition, immoderate desires, obstinacie in opinion, presumption, love of novelties, rebellion, disobedience."[86] Charron's debt to Cicero's Carneadean skepticism is obvious in these passages. The category of the probable (*pithanon*) was Carneades's specific contribution to the skepticism of the New Academy and Cicero promoted it in the Roman world in his own translation as *verisimile*.[87] In addition, Charron includes a direct quotation from Cicero's *Academica* II.8, where Cicero argues that holding views as merely "persuasive" (which is to "follow" rather than "affirm" them) is the great principle of intellectual freedom, since thus one avoids becoming "chained" to any view: "*hoc liberiores & solutiores sumus, quia integra nobis judicandi potestas manet*" ("but we are freer and less constrained because our power of judgment is intact").[88]

This is also the substance of Glanvill's response to White's condemnation of the *Vanity* as a piece of "Pyrrhonical contagion." Instead of a skepticism that blocks any kind of search, he declares himself an adept of the principle of the "freedom of inquiry," the safeguard against the "prepossession" and "vassalage" of the mind.[89] In this, he tells us, he takes his cue from others before him, such as Campanella, Bacon, as well as Avicenna and other ancient peripatetic philosophers, but his main model is Gassendi, who proposed principles "not in a confident and assertive form, but as probabilities and hypotheses."[90] What Glanvill has in mind here is an ideal rooted in Academic skepticism, the *libertas philosophandi,* of which Gassendi was an open and enthusiastic promoter and which he confessed he learned from his reading of Cicero and Charron.[91] The same ideal, understood more largely as freedom from authority, had a long line of early modern supporters, including, indeed,

[83] Charron 1606, 230–231.

[84] Ibid., 231.

[85] Ibid., 237.

[86] Ibid., 241.

[87] See Neto 2014, 32–34; Bellis 2017, 130, 144. Generally on Glanvill's skepticism, see Popkin 2003, 211–215; Van Leeuwen 1970, 71–87.

[88] Charron 1606, 241; Cicero 2006, 6 and Brittain's explanation (Cicero 2006, xxvi–xxvii); Neto 2014, 12, 22–23.

[89] Glanvill 1665b, 3, 8, 12.

[90] Ibid., 12, 13.

[91] Neto 2014, 46; Bellis 2017, 147–148.

Campanella and Bacon, as well as Descartes.[92] It was very likely on account of a strained inclusion of the (larger) idea of freedom from authority within the (narrower) idea of the Carneadean freedom of judgment that, in a further response to White, Glanvill made the odd suggestion that Bacon and Descartes were of a mind with Gassendi in "propos[ing] their Opinions as *Hypotheseis* [sic], that *may probably* be the true accounts, without peremptorily affirming that *they are*."[93]

The "probable" in its Carneadean-Ciceronian guise has a double aspect. One has to do with the epistemologico-psychological analysis of assent, which is at the heart of the debate between the stoics and the skeptics as recorded in the *Academica*. Since, contrary to the stoics, the skeptics do not think there are valid criteria of truth (such as the stoic "cataleptic impression"), they will refuse to embrace fully any impression, and thus will suspend assent. Yet, the Carneadean skeptic allows that some impressions are likelier than others (even if subsequently they will prove less likely than novel ones) and thus "follows" them, which is different from "affirming" them in any definite manner. This is therefore a thesis about the nature of the assent given to a likely impression. The other aspect has to do with a thesis about the indefinite temporality of inquiry. The assent given to a likely impression is considered provisional, since with further inquiry and examination, a likelier impression may present itself. Charron takes over the big lines of this argument, as does Glanvill, whose note – quoted above – about the "temporary determinations" of the true inquirer is an obvious echo of it.

Also like Charron, Glanvill uses the skeptical (and stoic) trope of the rashness of assent involved in the formation of erroneous belief.[94] In the section on the understanding in the *Essays* variant of *Vanity*, he rehearses the condemnation of what he variously calls the rashness, forwardness, precipitance or haste of assent. He also adds a positive passage that formulates the right method of inquiry in terms of the Ciceronian regulation of assent:

> [T]he *Understanding* must be *patient*, and *wary*, and *thoughtful* in seeking Truth; It must go step by step, and look every way, and regard many Things; It must distrust Appearances, and be shy of Assent, and consider again and again before it fixeth. This Method is necessary to the attainment of Knowledge [...].[95]

The idea had been there from the start, though: in both *Vanity* and *Scepsis*, the sections on the errors of the senses, especially of our childhood impressions, include passages that commend the same "method" (suspension of assent and impartial inquiry of appearances) – only that, there, this is praised as "the method of the most excellent Des-Cartes."[96] The corresponding passage in *Essays* adds the detail that makes *this* suspension of assent truly Cartesian: "to admit nothing but what we

[92] Sutton 1953; Stewart 1994; Maclean 2006.

[93] Glanvill 1676, 44. See further Cope 1956, 211–212; Talmor 1981, 13; Jesseph 2005, 191.

[94] Neto 2014, 24, 26.

[95] Glanvill 1676, 29. Glanvill is referring strictly to physical phenomena. For him, mathematical and divine truths are excepted from this suspension of belief: 1661, 209; 1665a, 152.

[96] Glanvill 1661, 73; 1665a, 55.

clearly, and *distinctly* perceive."[97] Again, Glanvill reads Descartes through Gassendist-Charronian (and thus Ciceronian-Carneadean) lenses, in virtue of his enthusiasm for an adept of the *libertas philosophandi*, however defined.

The regulation of assent, including its initial suspension and its cautious, provisional granting to what appears probable, is thus the first member of Glanvill's positive art of thinking, and it is rooted in the Academic-skeptical and anti-dogmatic idea of the freedom of the mind in inquiry. A second member, I suggest, is similarly derived and consists in a regulation of apprehension, answering to the idea of the careful examination of impressions. Together, these two instructions form the positive counterpart to the analysis of error, where medico-psychological material was grafted onto a late scholastic logical format. In addition, as we will see below, they generate the framework of the method of natural inquiry Glanvill promoted, and which won him the approbation of the Royal Society: the natural historical examination of instances and the discipline of the probable in theory-holding.

The best example of Glanvill's suggestion of a regulation of apprehension via natural history is one of the least 'modern' aspects of his thought: his interest in occult phenomena, apparitions and witches. *Vanity* and *Scepsis* record it under the title of "supposed impossibles," which is a section of the chapters where Glanvill assesses (and denies) the very possibility of dogmatic science, by means of an inquiry into whether a knowledge of causes is available to us. The *Essays* version drops this discussion altogether, and replaces it with the general statement that demonstrations are raised on "*Principles* of our *own*" (recalling Bacon's *analogia hominis* as a criterion of erroneous judgment) rather than on universal nature, or God's ways in nature; but these principles are variable, hence can only be granted the status of "*Hypotheses*, which *may be*, and *may be otherwise*."[98] The attack on 'demonstration' in the strong, scholastic sense of the term is indeed the central tenet of the chapters on the "supposed impossibles" in both *Vanity* and *Scepsis*, which treat of odd things such as sympathized hands and the magnetic cure of wounds (which are stock topics in natural magical literature). The argument is that a firm denial of the existence or even possibility of these and like phenomena proceeds from a dogmatic claim that grants certainty and universality to tenets which can only be partial and thus uncertain. Yet the fact that the natural magical material disappears from *Essays* does not mean that the mature Glanvill brushed aside his youthful extravagances (together with the flourishes of his early style)[99]; it cannot mean that, since *Saducismus Triumphatus*, which was published 6 years after the *Essays* is a much more extended discussion of precisely the same topic, even if illustrated with different (if equally extravagant) examples.

Glanvill asks: "Can nothing be otherwise, which we conceive *impossible*, to be so? [...] are not many things certain by the *Principles* of one, which are impossible

[97] Glanvill 1676, 22; cf. 1676, 46. For the Cartesian precipitance and suspension of assent, see Descartes 1984–5, II, 40–42.

[98] Glanvill 1676, 15. Bacon, *Novum organum* I.41, in Bacon 1996-, XI, 78.

[99] Medcalf 1970, xxi–xxii.

to the apprehensions of another?"[100] He adduces examples from natural philosophy (e.g., that the sun is in the center of our world), on a par with those invoked in the sections on the soul, the apprehension of which, he claimed, must be recognized as difficult (e.g., how the immaterial soul interacts with the material body), and continues the list with examples from natural magic: (a) that a person is able to bind the thoughts of another by the power of her imagination, (b) that two persons can communicate at a distance by means of magnetized needles, and (c) that wounds may be cured by applying medicine to the weapon that caused them. In order to argue for the probable status of the existence of such phenomena, Glanvill sketches several variants of an experimental methodology.

The binding of thoughts (the first example) is first instantiated with the famous story of the Gypsy scholar, as received from trustworthy "hands."[101] The next step is to list a series of other phenomena recorded by natural historical observation, such as the signature of the mother's imagination on the fetus, the transmission of laughter by sympathy, the unison of sympathized strings, and various kinds of fascination.[102] Further, the question of explanation is addressed, and two possible theories are offered – More's soul of the world and the mechanical account – with the proviso that both are to be entertained as "hypotheses" rather than infallible explanations.[103] In the case of the magnetized needles used for communication at a distance (the second example), Glanvill first describes the scenario of an experiment he urges the readers to undertake for themselves ("Let the friends that would communicate take each a Dyal," etc.).[104] The description employs fine detail of setting and operation and also comments on the effects (or lack thereof) of possible variables; it also closes with the expression of the hope that the success of the experiment will be greatly improved "when Magical History shall be enlarged by riper inspections."[105] Next comes the invocation of similar phenomena as reported by trustworthy testimony; in this case, the communication of thoughts by sympathized hands.[106] The magnetized needles and the sympathized hands are both instances of the same phenomenon, the "communication between the *phancies* of men," of which the first example (the binding of thoughts) was also a manifestation.[107] Therefore, the two

[100] Glanvill 1661, 193–194.

[101] Glanvill 1661, 196. Medcalf advances that the story of the Gypsy scholar already disappears from the *Scepsis* on account of the lack of any authority endorsing it (i.e., any authority comparable to Digby for the story of the sympathized hands – see note 106 below): Medcalf 1970, xxi–xxii. Cf. Cope 1956, 124–125. This does not sound right: true, the Gypsy scholar disappears, but the reference to "the judicious Naturalist my Lord *Bacon*" on the binding power of the imagination (Glanvill 1665a, 146) functions as a shortcut for it (with reference to the *Sylva Sylvarum*, century X) – and one that brings in no small authority.

[102] Glanvill 1661, 199; 1665a, 147.

[103] Glanvill 1661, 199–200; 1665a, 147–149.

[104] Glanvill 1661, 203; 1665a, 149.

[105] Glanvill 1661, 204; 1665a, 150.

[106] Glanvill 1661, 205–206; 1665a, 150–151.

[107] Glanvill 1661, 207; 1665a, 151.

sets of observational and experimental procedures are capable of corroborating each other. And they will be further reinforced by the third example, of the magnetic cure of wounds. Glanvill's approach here is the same as the explanatory step in the first example: the presentation of alternative explanations – either in terms of Digby's mechanism or by way of More's *anima mundi*.[108]

In sum, Glanvill's experimental method prescribes two main procedures. On the one hand, the use of varied instances of the phenomenon under investigation, the reliance on trustworthy testimony, and the framing of experiments with specified conditions and variables. These are members of the early modern method of experimental natural history, as embraced by the Royal Society and codified by Francis Bacon. Of particular interest is the variation of experiment, which was one of the key contributions of Bacon's art of experimentation (the *experientia literata*). Glanvill's use of multiple instances recalls the 'mode of experimentation' Bacon called *variatio* (experiment "tried in other things of a like kind"); moreover, Glanvill's parallel between natural-philosophical and natural-magical examples is possibly an echo of Bacon's *translatio* (the transfer of an experiment from one art to another, or from one part of an art to another part of the same art).[109] On the other hand, Glanvill's method also includes the formulation of theories that could make sense of the phenomenon; this will be a non-dogmatic consideration of alternative theories, held as probable, rather than as definitive explanatory accounts.

Let us recall that all of this came out of Glanvill's plea for a non-dogmatic freedom of mind in natural inquiry, which took the form of an art of thinking that consisted of two main heads, the regulation of apprehension and the regulation of assent. In turn, this art was a response to the analysis of error in terms of the mind's flawed operations, chief among which were the fixedness of a strong apprehension (whether perceptual or propositional) and the rushed movement of assent, resulting in the obstinate and partial stance of the dogmatic mind. The guiding scheme here was that of the three logical acts, on which psychological, medical-philosophical and skeptical material came to be grafted. Glanvill's art of thinking and by extension his method in natural inquiry are meant to counter both these tendencies: the natural historical variation of instances will shake the mind out of the blockage of overpowering apprehensions; and the discipline of holding explanatory theories as probable and provisional will temper the forwardness of the mind's assent.

This is also present in Glanvill's paean to the Royal Society, the *Plus Ultra* of 1668, where the natural historical method and the discipline of the probable in theory-holding appear in the form of an attack on "speculation," understood as the peremptory establishment of theories on too scarce an experimental basis:

> They continually declare against the *establishment* of *Theories*, and *Speculative Doctrines*, which they note as one of the most considerable *miscarriages* in the Philosophy of the Schools: [...] Since their *Aims* are to free *Philosophy* from the vain *Images* and *Compositions* of *Phansie*, by making it *palpable*, and bringing it down to the *plain objects* of the *Senses*;

[108] Glanvill 1661, 207–208; 1665a, 151–152.

[109] Bacon, *De augmentis*, in Bacon 1963–1994 (1857–1874), IV, 413, 416. 'Experiment' is meant here, following early modern usage, as both observation and intervention.

> For those are the *Faculties* which they employ and appeal to, and complain that *Knowledge* hath too long hover'd in the *clouds* of *Imagination.*[110]

Glanvill's own substantive contribution to this program – as I suggested above – took the form of natural histories of the non-palpable. The chapters on the "supposed impossibles" in *Vanity* and *Scepsis*, which treated of the occult phenomena of communication via the imagination, find a natural sequel in the concern of *Saducismus Triumphatus* to argue for the possibility of another set of "supposed impossibles," which included apparitions, ghosts and the extraordinary actions of witches; these are also often reliant on the natural-magical powers of the imagination.[111] Glanvill establishes the continuity between this account and that of the natural philosophical instances of difficult apprehension in *Vanity* and *Scepsis* (e.g., how the immaterial soul moves the material body)[112] and re-states his method: testimonies and "demonstrations of fact," which should trump the opinions and "*conceits* of our bold *imaginations*" that make us deny "*matters of fact* [...] because we cannot *conceive how* they can be performed"; and the cautious proposal of their "probable *causes*, shewing how things *may be*, not presuming how they *are*."[113] The ultimate benefit of this investigation for Glanvill is that it will argue for the existence of the immaterial realm, and thus will contribute to the defense of the immortality of the soul. This brings us full circle to where we (and Glanvill) began: the immortality of the soul as the framework for his inquiry into faulty and correct thinking, and by extension for his experimental method of natural inquiry.

9.5 Conclusion

Glanvill's initial concern with the immortality of the soul took an unexpected turn as it led him to consider the matter of how we come to the knowledge of immaterial things. Questions of knowledge had been traditionally asked either within the *de anima* part of a physics or under logic. Glanvill was thus naturally brought to investigate the matter from the point of view of these disciplines. But he lived in an age that fervently sought to redefine and rearrange disciplinary contents and boundaries, and was often eager to embrace eclecticism in doing so. We have seen how he used the resources of late scholasticism, humanism and the new philosophies, and melted them together in a compound that is far from original (nor was this necessarily a virtue in this period), yet the end-result of which suited the 'progressive' attitudes of the Royal Society. Glanvill was a man of his times and one of the larger outcomes

[110] Glanvill 1668, 89.

[111] Glanvill, "Some Considerations about Witchcraft," in Glanvill 1682, 10 (for the topic of "supposed impossibles"), 78 (for the natural magical imagination).

[112] Ibid., 11.

[113] Ibid., 12. Note that the imagination appears here in a twofold guise: the epistemic imagination can block the apprehension of phenomena attributable to the natural magical imagination.

of the discussion here is a glance at this feature of the early modern age through the lens of his often neglected writings. In his hands, physics and logic became overlapping territories, to the point where incongruities and tensions were bound to appear. Medicine was added to the mix, as were several choice bits of skepticism. In the process, the Schools, Charron, Bacon, Gassendi, Descartes and More were made to stand side by side, often more uneasily than they themselves would have allowed.

Yet despite the tensions, there is coherence to Glanvill's account. It seems to me that it all falls into place if we see what he is doing as driven by the idea of an art of thinking well. Logic increasingly came to be defined as such, both in late scholastic and in new philosophical quarters.[114] Glanvill did not write a logic (nor did he claim he did) but the use of the three acts of the intellect as the organizing scheme in his evaluation of the human mind's epistemic performance – not only openly in the chapters on the imagination but also implicitly in the rest of the *Vanity* texts – argues for a logical inspiration in what he was doing. The material he culled from medico-philosophical and skeptical accounts was such that it served or supplemented this guiding scheme. Particularly noteworthy here is the way Glanvill grafted the Academic skeptical ideal of intellectual freedom, and the epistemic-cum-psychological recommendations it upheld, onto the logical aim. It is also owing to this graft, as well as to the psychological and medical material, that the place of the analysis of error in this art of thinking becomes very different from what it was in a traditional logic. Error is no longer presented as a fallacy resulting from the misapplication of the correct rules of thinking; instead, the art of thinking appears as a solution to the problem of error. Therefore, the diagnosis of the way the cognitive faculties go wrong occupies the first place not only in the order of presentation but also in the conceptual order, since it identifies a condition to be mended in order to enable the mind's access to truth. Bacon's doctrine of the idols of the mind pointed in this direction, as did Gassendi's canons of deception, and Descartes' analysis of the prejudices acquired in early education. Glanvill, we have seen, was indebted to all three of them.

The pervasive presence of the imagination in Glanvill's account represents yet another departure from a traditional logic, which is nevertheless sanctioned, in various ways, by the eclectic material he used in his diagnosis of error. The original impetus behind highlighting the imagination seems to have been the platonic and neoplatonic indictment of this faculty, which served the purpose of paving the route to the issue of the immortality of the soul. Along the way, however, as the topic of right and wrong modes of thinking took precedence over that issue, other sources came to play their part: Gassendi's own grafting of physics and logic; the medico-theologico-philosophical literature on enthusiasm and melancholy; and Charron's Academic skeptical account of wisdom. Another larger outcome of the discussion here, therefore, is the inclusion of Glanvill's thought as a noteworthy chapter in the early modern history of the imagination, in particular of the role of the imagination

[114] Jardine 2007 (1988); Nuchelmans 1998, 104–105; for the specific case of Oxford, see Feingold 1997, 280–282.

in accounts of scientific method. The imagination does have a role to play in these accounts precisely because it becomes involved in the new arts of thinking, with which the question of method is tightly intertwined in this period.

The role the imagination openly plays is that of the faculty responsible, together with the body, for the various types of errant behavior of the mind. As a consequence, the recommendations for the right method of thinking and thus the right method of scientific inquiry seemingly take the form of operations that *reason* needs to perform *against* the imagination. However, the details of the analysis suggest another possibility, namely, that the regulation takes place *within* the territory of the imagination. It is not really the intellect that acts to block the bad effects of the imagination in Glanvill's account. On the one hand, even if the area of the action of the intellect is identified as that of non-sensible things, the right way to apprehend such things, he claims, still goes through the senses and the natural historical collection of instances of their material manifestations, just as in a natural historical investigation of sensible things. On the other, both the regulation of apprehension via the variation of instances and the regulation of assent via the discipline of the probable work on particular qualities of the acts *of* the imagination, which they seek to modify or temper: they contribute to the dissolution of the fixedness of the strong impression and to the moderation of the rushed movement of assent. In a sense, Glanvill's art of thinking may therefore be called a rational art of the imagination.

Except that, in the course of the analysis, the understanding comes back, not in the guise of the cognizer of non-material things, but as the faculty of assent, and when positive prescriptions are offered, it occupies center stage. This is very likely owing to the *evaluative* distinction between imagination and intellect that Glanvill introduces alongside the descriptive one; as well as to the fact that the strongest influence here is that of the skeptical and stoic analyses of assent, which attributed it to a rational (and unified) mind. The unclear division of labor between imagination and intellect appears as yet another tension in Glanvill's eclectic thought. However, there are potential benefits this tension can bring from the point of view of historical explanation. On the one hand, it can be seen as the result of a shift of focus from the *faculties* to the *acts* of the mind – a forward-looking development of a scholastic legacy. On the other, while the mental acts or operations acquire prominence in later works on the human mind (themselves understood by their authors as varieties of logic), faculty talk remains relevant if somewhat confusing. We can now begin to understand the context of this confusion. The tensions in Glanvill's account may indicate something about why it is that when Locke talks about the understanding and Hume about the imagination, they actually talk about largely the same things.[115]

[115]Compare Hume and Locke on complex ideas: Hume 1965, I.1.vi, 13; Locke 1975, II.xii.1–3, 163–164. For Locke, the complex ideas of modes, substances and relations are achieved by specific "acts of the mind" which he identifies as: combining into one compound, bringing together without uniting, and separating. Locke also notes that these acts belong to a "faculty of repeating and joining together its *Ideas*," whereby the mind "has great power in varying and multiplying the Objects of its Thoughts, infinitely beyond what *Sensation* and *Reflection* furnished it with". This power had

Bibliography

Anonymous [Henry More]. 1682. The true notion of a spirit. In *Saducismus triumphatus, or, full and plain evidence concerning witches and apparitions*, ed. Joseph Glanvill, 2nd ed. London: S. Lownds.

Ariew, Roger. 2014. *Descartes and the first Cartesians*. Oxford: Oxford University Press.

Bacon, Francis. 1963–94 (1857–74). *The works of Francis Bacon*, ed. J. Spedding, R.L. Ellis and D.D. Heath, 14 vols. Stuttgart-Bad Cannstatt: Frommann.

———. 1996. In *The Oxford Francis Bacon*, ed. Graham Rees et al., vol. 15. Oxford: Oxford University Press.

Bellis, Delphine. 2017. *Nos in diem vivimus*: Gassendi's probabilism and academic philosophy from day to day. In *Academic scepticism in the development of early modern philosophy*, ed. Plínio Junqueira Smith and Sébastien Charles, 125–152. Dordrecht: Springer.

Bernier, François. 1648. *Abrégé de la philosophie de Gassendi en VII tomes*. 2nd ed. Lyon: Anisson, Posuel et Rigaud.

Burns, William E. 2004. Glanvill, Joseph (1636–1680). *Oxford dictionary of national biography*. Oxford University Press. online edn., Jan 2008. http://www.oxforddnb.com.ezproxy.library.uq.edu.au/view/article/10790, Accessed 6 Apr 2017.

Burton, Robert. 1989–1990. *The anatomy of melancholy*, 3 vols., ed. Thomas C. Faulkner, Nicolas K. Kiessling, and Rhonda L. Blair. Oxford: Clarendon Press.

Casaubon, Meric. 1655. *A treatise concerning enthusiasme*. London: Thomas Johnson.

Charron, Pierre. 1606. *Of wisdome three bookes*. Trans. Samson Lennard. London: Eliot's Court Press.

Cicero. 2006. *On academic scepticism*, trans., introduction and notes Charles Brittain. Indianapolis/Cambridge: Hackett Publishing Company.

Collegium Conimbricense. 1600. *Commentarii collegii conimbricensis societatis Iesu, in tres libros de anima Aristotelis stagiritae*. Cologne: Lazarus Zetznerus.

Cope, Jackson. 1956. *Joseph Glanvill, Anglican apologist*. St. Louis: Washington University.

Corneanu, Sorana. 2017. The nature and care of the whole man: Francis Bacon and some late renaissance contexts. *Early Science and Medicine* 22: 130–156.

Corneanu, Sorana, and Koen Vermeir. 2012. Idols of the imagination: Francis Bacon on the imagination and the medicine of the mind. *Perspectives on Science* 20: 183–206.

Crignon-De Oliveira, Claire. 2008. Enthousiasme et sorcellerie chez Henry More et Joseph Glanvill: les limites de la critique de l'enthousiasme. *Revue de Métaphysique et de Morale* 59: 323–335.

Descartes, René. 1984–85. *The philosophical writings of Descartes*, 2 vols., ed. and trans. John Cottingham, Robert Stoothoff, and Dugald Murdoch. Cambridge: Cambridge University Press.

Digby, Kenelm. 1644. *Two treatises. In the one of which, the nature of bodies; in the other, the nature of mans soule; is looked into in the way of discovery, of the immortality of reasonable soules*. Paris: Gilles Blaizot.

Eustachius a Sancto Paulo. 1609. *Summa philosophiae quadripartita, de rebus dialecticis, ethicis, physicis, & metaphysicis*. Paris: C. Chastellain.

Feingold, Mordechai. 1997. The humanities. In *The history of the University of Oxford: Volume IV, seventeenth-century Oxford*, ed. Nicholas Tyacke, 211–358. Oxford: Oxford University Press.

Galen. 1821–33. *Claudii Galeni, Opera omnia*, ed. C. G. Kühn, 20 vols. in 22 parts. Leipzig: Knobloch.

Garber, Daniel. 1998. Soul and mind: Life and thought in the seventeenth century. In *The Cambridge history of seventeenth-century philosophy*, ed. Daniel Garber and Michael Ayers, vol. 1, 759–795. Cambridge: Cambridge University Press.

---

been ascribed to the imagination before and would be again, in Hume. For the relation between Locke's *Essay* and logic, see Serjeantson 2008; on Hume's *Treatise* and logic, see idem, 2005.

Gassendi, Pierre. 1658. *Opera omnia in sex tomos divisa*. Lyon: Anisson et Devent.

———. 1981. *Institutio logica 1658. A critical edition*, ed. and trans. Howard Jones. Assen: Van Gorcum.

Glanvill, Joseph. 1661. *The vanity of dogmatizing; or confidence in opinions, manifested in a discourse of the shortness and uncertainty of our knowledge, and its causes*. London: Henry Eversden.

———. 1665a. *Scepsis scientifica; or, Confest ignorance, the way to science; in an essay of the vanity of dogmatizing, and confident opinion*. London: Henry Eversden.

———. 1665b. *Scire/i tuum nihil est; or, The author's defence of The vanity of dogmatizing; Against the exceptions of the learned Tho. Albius in his late Sciri*. London: Henry Eversden.

———. 1668. *Plus ultra, or, the progress and advancement of knowledge since the days of Aristotle*.... London: James Collins.

———. 1676. *Essays on several important subjects in philosophy and religion*. London: John Baker.

———. 1682. *Saducismus triumphatus, or, full and plain evidence concerning witches and apparitions*. 2nd ed. London: S. Lownds.

Gowland, Angus. 2008. The ethics of renaissance melancholy. *Intellectual History Review* 18: 103–117.

———. 2012. Consolations for melancholy in renaissance humanism. *Society and Politics* 6: 10–38.

Harrison, Peter. 2007. *The fall of man and the foundations of science*. Cambridge: Cambridge University Press.

Heyd, Michael. 1995. *"Be sober and reasonable": The critique of enthusiasm in the seventeenth and early eighteenth centuries*. Leiden: Brill.

Hume, David. 1965 (1888). *A treatise of human nature*, ed. L.A. Selby-Bigge. Oxford: Clarendon Press.

Jardine, Lisa. 2007 (1988). Humanistic logic. In *The Cambridge history of renaissance philosophy*, ed. Charles B. Schmitt, Quentin Skinner, Eckhard Kessler, and Jill Kraye, 173–198. Cambridge: Cambridge University Press.

Jesseph, Douglas. 2005. Mechanism, skepticism, and witchcraft: More and Glanvill on the failures of the Cartesian philosophy. In *Receptions of Descartes: Cartesianism and anti-Cartesiansim in early modern Europe*, ed. Tad Schmaltz, 183–199. London/New York: Routledge.

Kessler, Eckard. 2007 (1988). The intellective soul. In *The Cambridge history of renaissance philosophy*, ed. Charles B. Schmitt, Quentin Skinner, Eckhard Kessler, and Jill Kraye, 485–534. Cambridge: Cambridge University Press.

Levitin, Dmitri. 2015. *Ancient wisdom in the age of the new science. Histories of philosophy in England, c. 1640–1700*. Cambridge: Cambridge University Press.

Lewis, Rhodri. 2006. 'Of Origenian Platonisme': Joseph Glanvill on the pre-existence of souls. *Huntington Library Quarterly* 69: 267–300.

Locke, John. 1975. *An essay concerning human understanding*, ed. Peter Nidditch. Oxford: Clarendon Press.

Lolordo, Antonia. 2007. *Pierre Gassendi and the birth of early modern philosophy*. Cambridge: Cambridge University Press.

Lupoli, A. 1984. 'Scetticismo moderato' e Aristotelismo antiscolastico: la polemica tra Joseph Glanvill e Thomas White. In *La storia della filosofia come sapere critico. Studi offerti a Mario dal Pra*, 122–155. Milano: Franco Angeli.

MacDonald, Paul S. 2013. Introduction. In *Kenelm Digby's Two treatises*, ed. Paul S. MacDonald, 5–38. Unspecified: The Gresham Press.

Maclean, Ian. 2006. The 'sceptical crisis' reconsidered: Galen, rational medicine and the *libertas philosophandi*. *Early Science and Medicine* 11: 247–274.

Neto, José R. Maia. 2014. *Academic skepticism in seventeenth-century French philosophy. The Charronian legacy 1601–1662*. Dordrecht: Springer.

Medcalf, Stephen. 1970. Introduction. In *The Vanity of dogmatizing: The three 'versions'*, xiii–xilv. Hove, Sussex: The Harvester Press.

More, Henry. 1656. *Enthusiasmus triumphatus; or a brief discourse of the nature, causes, kinds, and cure of enthusiasm*. London: James Flesher.

———. 1659. *Immortality of the soul, so farre forth as it is demonstrable from the knowledge of nature and the light of reason*. London: James Flesher.

Nuchelmans, Gabriel. 1998. Logic in the seventeenth century. In *The Cambridge history of seventeenth-century philosophy*, ed. Daniel Garber and Michael Ayers, 2 vols., vol. 1, 103–117. Cambridge: Cambridge University Press.

Park, Katharine. 1974. *The imagination in renaissance psychology*, M.Phil. diss. The Warburg Institute: University of London.

Pico della Mirandola, Gianfrancesco. 1501. *Liber de imaginatione*. Venice: Aldine Press.

Popkin, Richard. 2003. *The history of scepticism from Savonarola to Bayle,* new and expanded ed. Oxford: Oxford University Press.

Reynolds, Edward. 1640. *Treatise of the passions and faculties of the soule*. London: Robert Bostock.

Sanderson, Robert. 1615. *Logicae artis compendium*. Oxford: Iosephus Barnesius.

Savino, Christina. 2011. La ricezione del *Quod animi mores* di Galeno fra medioevo e rinascimento. *Bruniana & Campanelliana* 17: 49–64.

Serjeantson, Richard. 2005. Hume's general rules and the 'chief business of philosophers'. In *Impressions of Hume*, ed. Marina Frasca-Spada and P.J.E. Kail, 187–212. Oxford: Oxford University Press.

———. 2008. 'Human understanding' and the genre of Locke's *Essay*. *Intellectual History Review* 18: 157–171.

Shapiro, Barbara. 1983. *Probability and certainty in seventeenth-century England*. Princeton: Princeton University Press.

Spiller, Michael R.G. 1980. *'Concerning natural experimental philosophie': Meric Casaubon and the Royal Society*. The Hague/Boston/London: Martinus Nijhoff.

Stewart, M.A. 1994. *Libertas philosophandi*: From natural to speculative philosophy. *Australian Journal of Politics and History* 40: 29–46.

Sutton, Robert B. 1953. The phrase *libertas philosophandi*. *Journal of the History of Ideas* 14: 310–316.

Talmor, Sascha. 1981. *Glanvill: The uses and abuses of scepticism*. Oxford: Pergamon Press.

Van Leeuwen, Henry G. 1970. *The problem of certainty in English thought, 1630–1690*. 2nd ed. The Hague: Martinus Nijhoff.

Velcurio, Johannes. 1537. *In philosophiae naturalis partem omnium praestantissimam, hoc est, Aristotelis de anima libros, epitome longe doctissima*. Basel: Balthasar Lasius and Thomas Platter.

Vermeir, Koen. 2008. Imagination between physics and philosophy: On the central role of the imagination in the work of Henry More (1614–1687). *Intellectual History Review* 18: 119–137.

White, Thomas. 1646. *Institutionum peripateticarum ad mentem... K. Digbaei pars theorica*. London: Iacobi Rantonnet.

———. 1656. *Peripatetical institutions. In the way of that eminent person and excellent philosopher Sir Kenelm Digby*. London: John Williams.

———. 1665. *An exclusion of scepticks from all title of dispute: Being an answer to The vanity of dogmatizing*. London: John Williams.

Willis, Thomas. 1683. *Two discourses concerning the soul of brutes*. London: Thomas Dring.

Wright, Thomas. 1604. *The passions of the minde in generall*. London: Valentine Simmes and Adam Islip.

Chapter 10
Treating Yourself: Self-Diagnosis Amongst Natural Philosophers and Physicians and the Early Modern Medical Case Study

Anna Marie Roos

Abstract To test the parameters of the extent to which bodily health, bodily pain and the development of a uniquely natural philosophical understanding of nature were experienced and expressed in the late seventeenth century, this chapter will analyze three cases where early modern English physicians and natural philosophers made a diagnosis. But in these cases of John Wilkins (1614–1672), Martin Lister (1639–1712), and John Ray (1627–1705), their descriptions and diagnoses were of their own illnesses, their descriptions were not in a published treatise, but in private correspondence. As their own bodies were their dataset, examining their letters elucidates where datasets and mindsets meet in the seventeenth century. On the one hand, Wilkins, Lister, and Ray displayed clinical detachment and a high level of empirical detail about their own suffering. They were their own experimental subjects. On the other hand, they expressed emotional despair too along with data presentation, and their own pain was pre-Cartesian, a broken state of nature. One might argue that the private nature of their correspondence makes these sources problematic for analyses of public practice of natural philosophy, but in the case of Lister, his medical case was published anonymously in the *Philosophical Transactions of the Royal Society*, and reprinted in his *Letters and Mixt Discourses*. This incidence in itself assists us in elucidating norms for formulation and publication of early modern medical and natural philosophical case studies.

Keywords John Wilkins · Martin Lister · John Ray · Medical case studies · The Royal Society · Self-diagnosis · Pain · Dataset

---

Bodl = Bodleian Library, University of Oxford
RS = Royal Society Library, London
All translations are the author's own.

A. M. Roos (✉)
University of Lincoln, Lincoln, UK
e-mail: aroos@lincoln.ac.uk

G. Manning (ed.), *Testimonies: States of Mind and States of the Body in the Early Modern Period*, Archimedes 57,
https://doi.org/10.1007/978-3-030-39375-5_10

> English, which can express the thoughts of Hamlet and the tragedy of Lear, has no words for the shiver or the headache…. The meerest schoolgirl when she falls in love has Shakespeare or Keats to speak her mind for her, but let a sufferer try to describe a pain in his head to a doctor and language at once runs dry. There is nothing ready made for him. He is forced to coin words himself, and taking his pain in the hand, and a lump of pure sound in the other (as perhaps the people of Babel did in the beginning), so to crush them together that a brand new word in the end drops out. Probably it will be something laughable. – Virginia Woolf, *On Being Ill*[1]

In April 2015, I experienced my first bout with chronic pain, awakened in a hotel room in the middle of the night with my left side catatonically in spasm. My first thought was "I am going to die," and as the pain went from agonising to merely debilitating, I gathered myself together and went with my luggage to the Accident and Emergency Department at University College London Hospital. In the queue was one Greek gentleman bent over, twirling circles, shouting and gurning, and I realised he had a kidney stone wedged in his ureter. A young man behind me was in agony due to a tooth bleaching gone badly wrong, and another middle-aged woman was in a wheelchair with a swollen ankle, having as she described it "fallen into a hole." I waited, and waited some more, dragging my suitcase along in the queue. An hour later, I was ushered in to see the nurse who asked me triage questions to rule out appendicitis or internal bleeding. Urine and blood samples were taken, and I was dosed with Paracodol, a cocktail of codeine and paracetamol. There began another 3 months of painkillers (ibuprofen good, opiates bad) and ultrasounds, where I was mistakenly informed I had kidney stones, and then ovarian cysts, and at one point, even an ectopic pregnancy was mooted. My GP, frustrated that I was not getting any better, sent me to a neuromuscular clinic, where one wise physician finally said, "You must have had a severe muscle spasm of the left *quadratus lumborum* and now are suffering from chronic backache. Welcome to middle age, and do not slouch so much." Three months of osteopathic adjustments later, the pain is occasional, when it flares consisting of a tolerably dull ache, which is exacerbated by stress and overwork and alleviated by alcohol, sleep and mental distraction.

Why the misdiagnoses? Some of it was NHS bureaucracy, my case another burden on an already overburdened system, but some of it was due to the nature of pain itself. "Physical pain does not simply resist language but actively destroys it, bringing about an immediate reversion to a state anterior to language, to the sounds and cries a human being makes before language is learned."[2] The physician has a tricky job to hear the fragmentary and visceral knowledge of pain, coax it into clarity and interpret it into a patient history in order to make an accurate diagnosis.[3]

This chapter will show it was in the early modern period that natural philosophers and physicians started finding their voice to describe this visceral knowledge of pain, and they started with describing their own maladies. Self-diagnosis was a

---

[1] Woolf 2012, 6. This edition includes an essay by Woolf's mother, Julia Stephen, "Notes from Sick Rooms."

[2] Scarry 1985, 4.

[3] Ibid., 5.

logical place to begin in refining the patient history and case study, because it was a doubly tricky process in the early modern period. Other than chymical taste-assays, there were few "laboratory results, other than fairly unreliable uroscopy, to establish a disease as an objectified condition."[4] Furthermore, pain itself was conceived as a suffering of the soul, and not as a separate symptom. "The soul felt the injury to the body directly as pain, because body and soul, though terminologically distinct, were still imagined as one reality and could therefore not be separated."[5] Pain was a broken state of nature, and with limited availability in the early modern period to pain management, not something that could be fought or escaped, but only calmed, defied, denied, or accepted. While we tend to shun pain, early modern people "could find meaning in it as a mark of God's favor or his displeasure."[6] The patient could only be patched up; in the Hippocratic/Galenic tradition that dominated early modern physic, there was merely a continuum of health and illness, most people forever suspended between minor fluctuations of humours that meant total health was an unattainable ideal.[7] As Lindemann remarked, "even minor oscillations had to be dealt with expeditiously to avert illness."[8] With Cartesian mind-body dualism pain became more conceived as "a message from the body to the consciousness"; according to Descartes, the animal spirits conveyed pain messages that would "constrict the heart, agitate the liver… [entering] those nerves causing the external movements which bear witness to these, such as those that excite tears, or wrinkle the forehead and cheeks, or dispose the voice to cry."[9] Pain was thus transformed from a broken state of nature to "a protector of life."[10] With the possibility of pain's total elimination and a state of absolute health, it was no longer necessary for the individual to pay attention to his or her body so closely; that eventually became the job of the professional physician who had to develop precise diagnostic tools.

With the rise of Cartesianism in the latter part of the seventeenth century, pain and disease would gradually become less functionalist and more ontological. In the functionalist approach to disease, the illness was seen as "extant only within a specific organism" and attributed to an individual's bodily constitution or environment, often in the context of Galenic humoralism and the four qualities of hot, cold, wet and dry. In the functionalist view, the body, intertwined with the soul, was thought to have its own intelligence; when Faust tried to prick his finger to sign away his soul to the devil, his finger refused to bleed in an act of intelligent self-preservation. By the nineteenth century, disease became more and more objectified, resulting in a

[4] Ragland 2012; Duden, 1991, 87.

[5] Duden 1991, 87.

[6] Lindemann 2010, 9. I am not saying that there is no case in which modern patients choose to accept pain rather than treating it, but rather that the first choice is often to do what we can to avoid it or ameliorate it.

[7] Ibid., 13.

[8] Ibid.

[9] Descartes 1998, 163.

[10] Duden 1991, 88.

firmly ontological view of disease as an independently existing entity.[11] The body in this case became just an anatomically intricate machine acted upon by pathogens, cured with antibiotics, and subject to surgical procedures.

Early modern physicians were in the midst of this transformation of the relationship between body, soul, health and disease. They also were in the midst of a period which considered how best nature herself was to be studied and described. As has been elucidated by the work of scholars such as Hal Cook, many early modern physicians were practising natural philosophers, influenced by the rise of Baconian empiricism, in which the operations of nature were to be described precisely as 'matters of fact,' repeatedly verified, and sifted and gathered to find natural laws.[12] If there was little instrumentation to diagnose maladies, and physical pain resisted description, perhaps the tools of Baconian empiricism could be employed to describe pain more accurately, make more specific diagnoses, and eventually to create a precise taxonomy of disease.[13] In this manner, medicine was intimately connected with the cultivation of the scientific mind. As Jeffrey Boss stated, for Bacon, "to learn about the body requires clinical observation and recording, together with the comparison of bodies, experiments on living animals and attention to pathological changes… Bacon has no separation of medicine from natural science; his philosophy of medicine is his general philosophy of the advancement of knowledge."[14]

To test the parameters of the extent to which bodily health, bodily pain and the development of a uniquely scientific understanding of nature were experienced and expressed, this essay will analyze three cases where early modern English physicians and natural philosophers made a diagnosis. But in these cases of John Wilkins (1614–1672), Martin Lister (1639–1712), and John Ray (1627–1705), their descriptions and diagnoses were of their own illnesses, their descriptions were not in a published treatise, but in private correspondence. As their own bodies were their dataset, examining their letters elucidates where datasets and mindsets meet in the seventeenth century. On the one hand, Wilkins, Lister, and Ray displayed clinical detachment and a high level of empirical detail about their own suffering. They were their own experimental subjects. On the other hand, they expressed emotional despair too along with data presentation, and their own pain was pre-Cartesian, a broken state of nature. One might argue that the private nature of their correspondence makes these sources problematic for analyses of public scientific practice, but in the case of Lister, his medical case was published anonymously in the *Philosophical Transactions of the Royal Society,* and reprinted in his *Letters and Mixt Discourses.*[15] This incidence in itself assists us in elucidating norms for formulation and publication of early modern medical and scientific case studies.

---

[11] Lindemann 2010, 12.

[12] Cook 1996a, b.

[13] As a proviso to this statement, recently appeared at this time was the pusilogium of Santorio and the medical thermometer, fruits of empiricism that led to more accurate medical diagnoses, if not to treat pain. See Bigotti and Taylor 2017.

[14] Boss 1978.

[15] Anonymous [Martin Lister] 1672; Lister 1683, 54–56.

10.1 The Case of John Wilkins

On 19 November 1672, Robert Hooke wrote in his Diary: "Lord Bishop of Chester died about 9 in the morning of a suppression of the urine."[16] The next day, Hooke elaborated: "Dr [Walter] Needham brought in account of Lord Chester's having no stoppage in his uriters nor defect in the kidneys. There was only found 2 small stones in one kidney and some little gravel in one uriter but neither enough to stop the water. 'Twas believed his opiates and some other medicines killed him, there being no visible cause of his death, he died very quickly and with little pain, lament of all."

Hooke noted in his *Diary* that as Wilkins lay dying, a team of "Royal Society medical men" treated him at his bedside.[17] Their treatments were of two sorts. The first stemmed from Galenic medicine that advocated expulsion of evil humours, as bladder or kidney stones thought to be due to excess humours in the urine. In personalities that were hotter in nature, these humours would harden or bake into stones that obstructed the ureter or bladder causing bleeding and pain. Physician Jonathan Goddard thus recommended "Blisters of cantharide (Spanish fly) apply'd to the neck or feet or to the veins," a traditional Galenic cure designed to draw the evil humours out of the body. Other cures were inspired by the new chemical medicine introduced since Paracelsus shook the medical world, particularly by Sylvius de la Boë's (1614–1672) iatrochemistry. Sylvius's theories were predominant at the University of Leiden medical school, in which bad digestion could be rectified by treatment with balancing acids and alkalis. Health was, in Sylvius's view, dependent on normal effervescences that occurred in digestion and bodily metabolism.

"Pathological conditions were reduced to two different types: either an excess of acid, an *acrimonia acida*, or an excess of alkali, and *acrmonia lixiviosa*."[18] These disturbances in the blood could be corrected by direct "chemical interference"; as disease was based on excess acidity or alkalinity, therapy was reduced to a correction of the excess by drugs of an opposite nature.[19] For example, Sylvius himself favoured *sal volatile oleosum*, an alkaline salt that he believed neutralised acid, stimulated respiration, and dissolved thick phlegm.[20] Much in the same spirit, one of Wilkins's attending physicians prescribed *flegma acidum succini rectificatum cum sale tartari* [lead precipitate of amber refined with salt of tartar], its acidity thought effective to dissolve the chalky alkaline kidney stones.[21] During their attendance at his bedside Wilkins's physicians carefully observed doses and outcomes, and their treatment regime was not unique. As is well known, Hooke himself engaged in physic-taking, minutely recording in his diary "his own in take of measured doses

[16] Hooke 1935, 11.

[17] Ibid., 13; Jardine 2004, 226.

[18] Beukers 1999, 52.

[19] Ibid., 53.

[20] Ibid.

[21] Hooke 1935, 13; Jardine 2004, 228.

of medicaments and their outcomes," making himself an experimental subject.[22] And, typical to the practice of early modern medicine for more elite patients, the worse the patient became, the more exotic the medicaments became; in a case of what can only be termed "iatrogenic regicide," King Charles II was dosed with over 40 different types of medicine including bezoar stone, extract of human skull, and finally pearl julep [a heart tonic] forced down the royal patient's throat. Charles jokingly apologized for "taking so long a dying," and Wilkins, ever the pragmatist, "say'd he was prepared for the Great Experiment."[23]

But 6 months before he died, Wilkins was not preparing for death, but instead penned a series of letters to the young physician Martin Lister concerning his illness in hopes of a cure. On 22 May 1672, he wrote:

> Sir,
> Though I am a stranger to your person, yet I am not so to your Worth nor to severall of your speciall friends. particularly to Mr Willoughby and Mr Ray, with whom I have lately discoursed concerning you. It is upon this Account, that I doe crave your advice. I did purpose this summer to have bestowed some time at Scarborough Spaw, having ben frequently troubled, with fitts of the stone, I doe not thinke that I have in my kidneys any formed stone. But some parcells of concreted sand, about the length of a graine of wheat which will easily crumble into powdere. ~~which~~ \And these/ in their passage through the Ureters, doe must trouble me. I had often heard the Scarborough Spaw much commended for such cases. But having lately seen a discourse by one Dr Tonstall, which asserts the contrary, and pretends to severall substances, I could not think of any better way for impartiall satisfaction, then by consulting your Opinion in the matter. I doe make it my request to you, that you will let me know what you think. Your letter (if the Post be payd forward to London) will be brought to me to Chester. If I should come into your parts, I shall hope to be better acquainted with you. for the meane space I begg your pardon for the trouble. from
> Sir your friend and servant.
> Joh: Cestriense[24]

Wilkins's desire for a medical opinion by post was nothing unusual. Medical diagnosis did not have to make use of physical examination of the patient, and physical examination could be secondary to a patient's verbal or written report. The top medical practitioner in eighteenth-century Edinburgh, William Cullen, routinely diagnosed his patients by post.[25] So even though the two gentlemen would never meet, Wilkins, learning of Lister's reputation through the recommendations of naturalists Francis Willughby and John Ray, became his patient. Wilkins and Ray had worked closely together, Ray making a Latin translation of Wilkins's *An Essay towards a Real Character and a Philosophical Language* in 1668. Ray also assisted Wilkins with his derivation of the language and attempted to apply it to his botanical taxonomy, and Willughby had assisted with his observations on birds. As for Lister, he was becoming known as an expert in the chemistry of spa waters, and would go on

---

[22] Hooke 1935, 13; Jardine 2004, 230. For further discussion of Robert Hooke, see the chapter by Felicity Henderson in this volume.

[23] Lamont-Brown 2001, 44–45; Shapiro 1969, 1.

[24] John Cestriense is Joannes Cestriensis or John of Chester, referring to his residence in the see of Chester. The letter may be found in Bodl. MS Lister 3, fols 193–194; Roos 2015, letter 210, 496

[25] Keller 2003, 69.

to publish a work on the English spa (*De fontibus medicatis Angliae exercitatio*, 1684). In his letter to Lister, Wilkins was also taking into consideration the public dispute about the medical efficacy of Scarborough Spa. George Tunstall of Newcastle, to whom Wilkins referred, served as Scarborough's physician from 1660 to 1664, but subsequently promoted spa waters at Harrogate. Tunstall was a chief antagonist of Robert Wittie, who wrote about the health benefits of Scarborough, commending it for everything from wind to leprosy.[26] Wilkins apparently wanted from Lister an independent and informed opinion, some "impartiall satisfaction" before taking the waters, and he appears to have been carefully weighing the received evidence. Wilkins also was quite correct that he had gravel in his ureters, which was subsequently confirmed *post-mortem* 2 years later.

Even though he must have been in considerable distress, kidney stones producing agonising abdominal pain, nausea, fever, and bloody urine, the only indication of Wilkins's distress was that he "craved" Lister's advice and he was "much troubled" by his malady. Perhaps his pain was so intense that it was indescribable, the occurrence of kidney stones frequent enough in early modern medical practice that words were not possible or necessary. Preventing kidney stones, gallstones or bladder stones in the body was a topic of great interest to early modern physicians, as medications were ineffective, and the operation to remove the stone was horrific. Claire Tomalin in her biography of the diarist Samuel Pepys described the procedure for bladderstones:

> The surgeon got to work. First he inserted a thin silver instrument, the *itinerarium*, through the penis into the bladder to help position the stone. Then he made the incision, about three inches long and a finger's breadth from the line running between scrotum and anus, and into the neck of the bladder, or just below it. The patient's face was sponged as the incision was made. The stone was sought, found and grasped with pincers; the more speedily it could be got out the better. Once out, the wound was not stitched—it was thought best to let it drain and cicatrize itself—but simply washed and covered with a dressing, or even kept open at first with a small roll of soft cloth known as a tent, dipped in egg white. A plaster of egg yolk, rose vinegar and anointing oils was then applied.[27]

This was the Marian method, named after Mariano, a Renaissance Italian physician. Trauma to the bladder neck and prostate was intense, there was usually severe hemorrhaging, and it was difficult to extract large stones through the male urethra.[28] Incontinence, erectile dysfunction, if not death, frequently resulted. It was not until Jacques de Beaulieu (1651–1714) developed a procedure that led to the modern lateral or perineal surgery that survival rates increased.[29]

---

[26] Wittie 1660; Simpson 1669. Simpson thought disease was due to faulty chemical reactions due to incorrect fermentation during digestion, and the minerals in the spa waters corrected this condition.

[27] Tomalin 2002, 63–4.

[28] Herr 2008 1215. Stones "would pass fairly easily through the short, dilatable female urethra." See Shelley 1958, 52.

[29] Pepys was lucky—his bladder stone, which was the size of a tennis ball (2 inches), came out intact, and the proud survivor had it set in a case that cost him 24 shillings.

One could also object that some of the formality of Wilkins's language in his letter may have been because Lister was a stranger; however, Wendy Churchill has demonstrated the emotive character of correspondence between early modern patients and doctors in cases of diagnoses by post, patients often expressing considerable anxiety about their illness.[30] But the overall impression is one of clinical detachment; Wilkins's description of his malady was predicated on treating his own body as an alien other, like a physician treats a patient. He could not control the pain the somatic body was creating, but he could control his reaction to it, and studying it clinically and analytically was both scientific and a coping mechanism. Though he did not know it conclusively, Wilkins's approach was an effective one. Recent studies have indeed provided the first "unambiguous evidence that 'affect labeling,'" or the explicit articulation of a negative emotion or experience like present or past pain, produces diminished responses to negative emotional images in the amygdala and other limbic systems and ultimately contributes to better health.[31] And so after describing his malady, Wilkins could then, like his fellow natural philosopher Robert Hooke, self-experiment giving himself a dose, observe the reaction to test its effects, and proceed.

And so he did, writing Lister again on 11 July 1672:

> … And this morning I waked with a fitt of the stone upon me which continued till about Eleven of the Clock. I am very sensible of your kindnes in being concerned for me. for the first ~~pre~~ letter you sent me, by which I was encouraged to come to this place, and for the second of advice about using the Waterrs.[32] They have not the same operation with me as with others, working chiefly by stoole and by salivation and very litle by Urine, which I principally hoped for. I have had two fitts of the stone since my coming and none for two or three month before. I begin to Question whether they may be proper in my Case.[33]

The high iron oxide or chalybeate content in the spa waters indeed meant they were "aparient" or served as a potent laxative, though they did have the reputation of being diuretic which unfortunately was not the case for Wilkins. As his experiment with the spa waters had not worked, Wilkins wrote Lister for the last time 4 days later, noting:

> I agree with you, that I ought not obstinately to contest against these waters, and therfore resolve to leave them this weeke. I am now going with Sir Hugh Cholmeley[34] to Whitby with a purpose of returning to morrow. I have a great desire to see you. But cannot at this time hope for it, by reason of the mishap, and my short stay in these parts. I thinke of going home on Ffriday."[35]

---

[30] Churchill 2013, *passim*.

[31] Lieberman, et al. 2007, 427.

[32] These letters have been lost.

[33] Bodl. MS Lister 3, fols 195–196; Roos 2015, letter 216, 464.

[34] Sir Hugh Cholmeley, fourth Baronet (1632–1689), politician and peer, of Fyling Hall, near Whitby in Yorkshire.

[35] Bodl. MS Lister 3, fols 196–197; Roos 2015, letter 217, 465.

Lister was also practising in York as a physician, so there was the possibility of a personal visit, but Wilkins's ill health and business prevented it, so despite the intimacy of their correspondence, the two never formally met.

Wilkins's self-treatment ineffective, the circumstances of his subsequent death were still a matter of continued debate decades later, subject to clinical analysis, observation and experimentation; in other words, retrospective diagnosis. Edward Baynard (1641–1719), a graduate of the University of Leiden, fellow of the Royal College of Physicians, and proponent of Sylvius's acid-alkali iatrochemistry, was convinced that Wilkins was misdiagnosed. In 1695, Baynard wrote in an article for *Philosophical Transactions* about a servant who laboured for 8 days with a suppression of urine, "the same Case, of which that most Learned Prelate, Dr Wilkins, late Bishop of Chester, dyed."[36] Baynard cured the servant by having him take a "quantity of Acids in a convenient Vehicle, upon which Secretion being presently made, he immediately urined in great quantity, and was thereby restored to his Health."[37] Rather than oyster shells or cantharides, Baynard believed acids dissolved kidney stones allowing for urination and should have been used in Wilkins's case. In Baynard's view, the colleague at Wilkins's deathbed who prescribed the acidic *flegma acidum succini rectificatum cum sale tartari* [lead precipitate of amber refined with salt of tartar], was offering the correct medicine. Modern medical historians are chary of retrodiagnoses based upon application of "correct" modern knowledge to disorders experienced in different cultural and social contexts. However, the new experimental thrust of early modern medicine and increasing emphases on case studies meant that seventeenth-century doctors themselves considered retrodiagnosis a possible and valuable means of investigation. It was a Baconian natural history to be analyzed, a piece of data recorded as part of the quest for understanding medical principles of nature.

10.2 The Case of Martin Lister

When Wilkins was writing Lister about his problems with the stone, Lister had just himself recovered from an eight-year malady involving another stone, this one not in his kidneys but under his tongue. He had contracted it in 1663 when embarking on a sea voyage to France to study medicine in Montpellier; Lister was caught in a storm in the channel crossing and stranded on Guernsey for a number of weeks, "wherein he suffered an exceeding cold; and that, not long after his landing, he found a certain Nodus or hard lump."[38] For the first 7 years, he was asymptomatic except when the weather turned cold.

[36] Baynard 1695, 20.

[37] Ibid.

[38] Anonymous [Martin Lister] 1672, 4062. The voyage was undertaken on 11 August 1663. The notes of his journey made in a pocket book (Bodl. MS Lister 19) indicate that his channel crossing was severely delayed by storms. Lister was stranded at Chesil Beach for 3 weeks and, after another

Most likely he was suffering from a blockage in the submandibular salivary gland, or Warthin's duct, which runs from under the front of the tongue to the submandibular gland. When saliva cannot enter a blocked or partially blocked duct, it backs up into the gland, causing pain and swelling in the gland. Pain is especially sudden and intense after starting a meal, and will ease 1–2 h after eating. If the gland is partially blocked, pain is intermittent which seems to have been the case for Lister for most of the duration of the disorder. Lister first mentioned the blockage to Royal Society secretary Henry Oldenburg on 13 September 1671, pleading illness to explain his late delivery of botanical specimens:

> You may expect some [plant] sprigs from me, as soon as I am able to goe abroad; for I have been very dangerously ill and the last Friday had an incision made under my Tongue, whence my surgeon tooke a stone out as bigg as a bean. I could be content to gratify your curiosity with the circumstances of this odd distemper, upon promise it may \it be/ not be published by the presse.[39]

In his typical investigatory fashion, Lister agreed the account of his 8-year long suffering from his tongue-stone could be published in the *Philosophical Transactions* as a case study for the benefit of his fellow virtuosi, as long as he was not identified as the patient.[40] Using case studies was a tried and true convention among early modern English doctors and surgeons writing medical tracts to demonstrate theoretical principles or the efficacy of a given prescription or surgical procedure.[41] However, Lister further distanced himself from his malady by routing his submission through two intermediaries, the Royal Society indicating in their minute book:

> Mr Oldenburg read a letter from Mr Lister to the Archbishop of York, dated at York 10th April 1672, and by his grace communicated to the lord Archbishop of Canterbury, who delivered it to Sir Robert Moray, in order to be imparted to the Society. It contained an exact account of the excision of a stone from under the tongue of a man, where it had been breeding about eight years. It had visible impressions upon it of some capillary vessels amongst which it grew.[42]

An anonymous publication in itself was not unusual or pejorative, and the early *Philosophical Transactions* were full of anonymous extracts; some contributors adhered to earlier norms of publication that had stressed that it was not gentlemanly to publish unless it was anonymous, as modesty should be paramount. Lister evinced these qualities when he visited London in 1670 and he met Oldenburg at home,

---

aborted crossing stranding him in Guernsey, he did not reach France until 16 October 1663. For an scholarly edition of MS Lister 19, see Anna Marie Roos, "Every Man's Companion, or An Useful Pocketbook: The Travel Journal of Dr. Martin Lister (1639–1712)," http://lister.history.ox.ac.uk/

[39] Letter from Martin Lister to Henry Oldenburg, 13 September 1671, RS EL/L5/38, printed in Oldenburg 1968–1971, vol. 8, 263–264, letter 1787, and Roos 2015, letter 183, 387–389.

[40] Anonymous [Martin Lister] 1672, 4062–4064

[41] Sandassie 2008, 11–18. For a general overview of the case history, see Riha 1995. I use the word 'case study' recognizing that they come out of the medieval consilia tradition, and that in early modern 'case studies' and consilia the identity of the physician is more important than in modern cases. See Pomata 2005.

[42] Birch 1756–1757, vol. 3, 48. The Archbishop of York was Richard Stearne (1596–1683).

without revealing himself as the author of another anonymous paper published 2 years earlier.[43] In this particular case, as was often so with medical case studies, the contributor to the *Transactions* wished to be anonymous to preserve his own identity or that of his patient(s).[44] In cases of less socially reputable diseases, such as STDs, even identifying some of the patients in case studies as being female could result in public criticism. Later in Lister's career he was roundly criticized by some of his peers in the Royal College of Physicians for a work on syphilis he wrote which contained case studies. Edward Lhwyd informed him, "Dr. Sherard and others had inform'd me that your book had given much offense to the London physicians: and I heard a young physician of great character in this town disparaging it, for some of the patients being women."[45] The vast majority of early modern medical writers supported a fermentation theory of syphilis in which the disease putrefied in the woman's womb, their case studies normally represented victimized men infected by women.[46] Moreover, early modern physicians writing about syphilis "often took this opportunity to pass moral judgment on female patients," and advocated the only safe sex could be with a "cold" or phlegmatic women whose constitution was less likely to lead to humoral fermentation and disease.[47] Lister, however, saw syphilis as gender neutral in his descriptions of culpable parties. Rather than concentrating on promiscuous women or ill-bred wet nurses who passed the disease to innocent husbands or children, in most of his case studies he described the devastating effects of a husband who infected his wife and children.[48]

Thus, as in Wilkins's case, Lister's actions of distancing himself from his own illness could have been a means of "affect labeling" to reduce emotional distress, but he also followed the norms of anonymity and clinical detachment required of published case studies. In his letter to the Archbishop of York and the subsequently published article, Lister wrote of himself in the third person:

> This extremity lasted 5 days, in all which time, the party had so vast a flux of spittle run[n] ing from him, that is was not possible for him to repose his head to sleep, without wetting all the bed about him; inso much as that is was very much questioned by some friendly visitants, whether he had not of himself, or by mistake, made use of some Mercurial medicine.[49]

He then went on to describe "the varieties or degrees of this spontaneous salivation" in vast detail, concluding with an account of the incision made by the surgeon to excise the stone. The stone was then donated to the Royal Society's Repository, its

[43] Hall 1999, 212. Lister only himself mentions the visit himself in a letter to Oldenburg on 9 August 1670. See Oldenburg 1968–1971, vol. 7, 104.

[44] Kronick 1988 232.

[45] Bodl. MS Lister 3, f. 132r. Sherard was James Sherard (1666–1738), an apothecary, and one of the most eminent botanists of his day. See Webb 2004.

[46] Siena 1998, 558.

[47] Ibid.

[48] For more information about Lister and case studies involving women, see Roos 2011, 359–60.

[49] Anonymous [Martin Lister] 1672, 4063. Mercury was a common Paracelsian cure for syphilis, a side effect excessive salivation.

first museum of objects from the natural world, again a not atypical action. The Royal Society Repository, as well as museums like the Ashmolean Museum had numerous donations of bodily stones as specimens, relics of illness and scientific inquiry.

But Lister went beyond displaying his relics, using his own and other's specimens to inform an iatrochemical investigation he was undertaking into the causes of what he called "petrifications" in the body. He invented a fairly elaborate theory that fool's gold (iron sulphides (FeS_2) or iron pyrites composed most bodily stones. His claim was a large departure from both Galenic medicine, which stated stones were a form of hardened phlegm, and from Paracelsianism, which saw them as akin to the passive sedimentation of the tartar of wine in the wine cask. Lister's rejection of Galen was common among other chemically-oriented physicians such as Frederick Slare (1647–1727), who stated in a paper he read before the Royal Society in 1683: "Nor is it necessary to derive the materiall Cause from such a Shiny and Roapy or Mucilaginous Indisposition of the humours, that may perhaps coagulate and harden into a Stone. Such a Viscous Urine being less apt to precipitate this gritty matter then more thin and limpid Urine."[50] Skepticism about the stone's being tartar also grew when gall and kidney stones were calcined and chemically analyzed as part of a series of experiments done on *materia medica* by the Royal Society in the 1670s and 1680s by physicians such as Slare and Nehemiah Grew, now keeper of the Royal Society Repository. Simply, Slare and Grew discovered that the specific gravity and reactions to acid and alkalis differed between tartar and bodily stones.[51] Grew studied medicine in Leiden under the tutelage of Sylvius, his dissertation concerning acid-alkali iatrochemistry and the distillation of bodily fluids, so it is not entirely surprising he would employ the same analytical methods to test gallstones.[52]

Lister's claims about "the stone," however, went even further, as they were a refutation of Van Helmont's treatise on gallstones, *De Lithiasi* (1648).[53] Van Helmont replaced the Paracelsian idea of tartar with a seminal fermentation and chemical transmutation of urine. The formation of the stone involved two agents or spirits—the spirit of wine and the spirit of urine producing a substance he called *duelech,* as it was dependent on the interaction of two different substances. This fermentation also operated independently of bodily heat or cold, and Galenic physicians saw it as important in the concoction of the stone and other bodily fluids. When Van Helmont attempted to recreate the formation of the stone in vitro, he mixed the volatile salt of urine with alcohol, obtaining a white precipitate which he called an offa, (a morsel or cake).

Lister pointedly rejected Helmont's theories concerning offa, stating "the white pellet Helmontius mentions… is no explanation of petrifaction, which is usually

[50] Slare, RBO.6.1, f. 1v. This paper is also printed as Slare 1683, 523.

[51] Slare, RBO.6.1, f. 3v.

[52] Grew 1675.

[53] van Helmont 1648.

accomplished without any urine-like spirit... from nothing other than... vitriol."[54] Why vitriol, and what did this have to do with pyrites and their powers? Scratching pyrites resulted in a sharp sulphurous smell, which according to the German mining theory of *witterung* promoted by authors such as Johann Rudolph Glauber (1604–1670) and Georges Agricola (1494–1555), was associated with fertile mines. The ore exhalations themselves were thought to contribute to the formation of minerals, stones and the transformation of matter. Left to themselves in a slightly damp environment, pyrites also produce a green vitriolic varnish and sometimes, spectacular green crystals of vitriolic salt (iron (II) sulphate). The pyrites' production of pungent smells and salts seemingly from thin air led many early modern natural philosophers to believe that pyrites were the source of a universal salt which could take on different forms according to the metals and minerals with which it came into contact.[55]

Lister thus proposed that this universal salt originating in pyrites could combine with intermediary substances in the body, forming, bladder and gall stones. Citing Galen, who claimed it was "impossible for bodies with a solid and compact substance, like stones, to decay", Lister claimed that since vitriol was stone-like in nature, they were not susceptible to "decay" or digestion in the stomach.[56] He was present at a Royal Society meeting in 1681 in which a paper was read claiming there were "five or six small stones shining like marcasites [a form of pyrite] taken out of the bladder of an ox". Lister apparently affirmed these finding, as he observed the same phenomenon, taking 150 shiny bladder stones out of an ox and sending them to the Royal Society for its repository. Lister then claimed that if bladder stones were made of pyrites, then human urine should also have pyrites in it. In support of his claim, he observed that his own urine and that of others deeply coloured silver chamber pots gold like the color of pyrites. (most likely due to the urea and bile salts present in the urine).

Lister then reasoned that a kind of magnetic force best explains the cause of petrifaction, as he believed petrified materials, which had as their source iron pyrites, were magnetic in nature. Lister presented a box of iron ores, pyrites and petrified wood to the Royal Society in a 1683 meeting, all of which were "examined by a good magnet". The next step was to see if human stones themselves were magnetic as pyrites, and as I have discussed elsewhere, his tests with bodily stones at Royal Society meetings met with seemingly positive results.[57] After his evidence in vitro that bodily stones were magnetic in nature, Lister next offered in evidence in vivo. In 1684, he presented an account to the Royal Society accompanied by a drawing he had commissioned of a "Stone Grown to an Iron Bodkin in the Bladder of a Boy,"

[54] Lister 1684, part II, 81–82: *Offa autem alba Helmontii... et minime quidem petrificationem explicat, quae sine ullo, aut spiritu urinoso... ex solo saxo calcario aut vitriolo abortivo.*

[55] Roos 2008, 147–51.

[56] Lister 1864, part II, p. 72.

[57] Roos 2011, chapter nine, *passim.*

and the report was published in the Philosophical Transactions later that year.[58] Apparently a young Parisian child had inserted an iron nail or needle in his urethra, it had made its way into his bladder, and a bladder stone had grown around it. After the lad was cut for the stone, the specimen was presented to the French king for his cabinet of curiosities. Lister, as did meeting participants Christopher Wren and Tancred Robinson, attributed the "sudden growth of the stone to the magnetism of the iron."[59]

Lister later reproduced this phenomenon by wrapping an iron nail in a cow's bladder and immersing it in "Hungarian vitriol" [usually iron (II) sulfate ($FeSO_4$) but also used for copper sulfate ($CuSO_4$)].[60] After a period of time, "facing the nail but on the outside of the bladder, some pure, bright copper adhered, in balls ranging from fairly large to very small, giving the appearance of being arranged like a chain, as far as the nail extended."[61] Lister concluded that: in this experiment the following singular and specific conclusions were manifest, that the copper was attracted and separated from its own salt through the magnetic influence of the iron, whilst the physical barrier of the bladder failed to prevent the collection and combination of the copper.[62]

Lister thus thought that he had experimentally verified that magnetism was behind the chemical formation of bodily stones, his investigations not only a result of addressing a common early modern medical problem, but completed as a result of his own suffering. Although his case study of his own illness was clinically detached and anonymous, following established norms, his own bodily excretions were fair game for chemical analysis and experimentation.

This delineation between privacy for case studies for illnesses contained within the body, and public demonstration of excretions outside the body may represent a transformation of the relationship between body, soul, health and disease, but also a shift in behavioral and bodily regulations that were tied to gender. According to Peter Stallybrass, in the sixteenth and early seventeenth centuries, men were defined in terms of the classical Galenic body, the image of a finished, completed man of closed individuality; women, on the other hand, were the "grotesques" that were unfinished, outgrew themselves, and transgressed their own limits.[63] Women by

[58] Lister 1685, 882.

[59] Birch 1756–1757, vol. 4, p. 357.

[60] Ecklund 1975, s.v. "hungarian vitriol." http://dbhs.wvusd.k12.ca.us/ChemHistory/ObsoleteChemTermsTOC.html.

[61] Lister 1711, 348. Although this work was written later, it summarizes Lister's research of petrifaction in the 1680s: *Verum illud singulare exemplum de virtute ferri magnetica, a me olim vulgatum, hic non tacebo; nempe virtiolum nativum Hunagricum aqua dissolvi; cui clavum majusculum praelongum, vesica bubula inclusum, immisi; processu temporis, contra clavum, vesicae bubulae extrinsecus adhaerebat purum putum cuprum, majusculis et minimis globulis.*

[62] Lister 1711, 348: *In hoc autem experimento illa singularia et specifica manifesta sunt; cuprum esse attractum, et a suo sale divisum, aura magnetica ferri, vesicae interpositione nihil accretioni et cohaerentiae cupri prohibente.*

[63] Stallybrass 1986, 124.

contrast were leaky—leaking menstrual blood, sexual lubricant, yeast infections, and lochial discharge after giving birth. The female body emphasized those corporeal parts that were open to the outside world such as the open mouth or genitals, but the masculine classical body was bounded, and associated with the head, the seat of reason. Additionally, the female grotesque body's favoured space was public, the marketplace, a place of disorder, its language that of festive obscenity and abuse, whereas the language of the male classical body was the tongue of official literature or of the ruling classes, a language governed by the etiquette of palaces, formal rhetoric, institutions, and private homes. Stallybrass claimed that because the female body was open, unlimited, and associated with disorder, it had to be put under surveillance, chastity, speech, and travel for women restricted.[64]

However, the medical case study by the seventeenth century moved beyond these gendered differences, its male physicians and virtuosi using the principles of Baconian empiricism to describing all patients in dispassionate, reasoned, objectified terms. Although, at first, social norms meant that case studies were necessarily anonymous, as a firmly ontological view of disease as an independently existing entity began to take hold, individuality ceased to matter anyway. An incidence in a disease, in "any-body," became just a data point, and the individual persona faded, only described by age and gender. So did the use of bodily excretions and substances; hence in the twentieth century, immortal lines of cancer cells were only identified as He-La, their donor Henrietta Lacks forgotten.[65] So, when Lister identified a syphilis patient as female, and was roundly chastised for doing so, he was at the cusp of this change; for him, a case study of venereal disease transcended gender norms. And, this was also the case with attitudes towards bodily secretions; although they had previously been considered feminine and disorderly, they were beginning not to be subject to gendered social regulation, but rather were materials to be quantified, distilled, displayed and discussed.

10.3 The Cases of John Ray

This state of affairs was even the case with sperm production, the most male and one of the most private of all bodily secretions, intimately tied with social and personal identity. It is commonly known that Dutch natural philosopher Antoine van Leeuwenhoek (1632–1723) used his own sperm in his microscopic studies, his observations a refutation of the doctrine of spontaneous generation, subsequently published in the *Philosophical Transactions*.[66] He did this at the urging of Henry Oldenburg, although as Ruestow described, Leeuwenhoek initially was "personally repelled by the subject," questioning the propriety of writing about semen and

[64] Ibid., 126.

[65] Skloot 2010.

[66] Ruestow 2004, 216.

coitus.[67] But in 1677, Johan Ham, a student from the medical school at Leiden, brought him a specimen ostensibly of semen from a gonorrhea patient in which Ham himself had found "living animals, which Leeuwenhoek now observed as well."[68] Leeuwenhoek resumed his experimental work, observing a vast numbers of "animalcules," which he likened to "small earth-nut[s] with a long tail[s]."[69] In a letter of 1677 to Lord William Brouncker, the President of the Royal Society, he reported his findings:

> What I investigate is only what, without sinfully defiling myself, remains as a residue after conjugal coitus. And if your Lordship should consider that these observations may disgust or scandalise the learned, I earnestly beg your Lordship to regard them as private and to publish or destroy them as your Lordship sees fit.[70]

The work was eventually published nearly 2 years later, in Latin, unlike most of his other works that were translated from the Dutch to English by Oldenburg. This may have been for a reason of propriety, but more likely it was due to Oldenburg's sudden demise in September 1677 and the ensuing chaos with the publication of the *Philosophical Transactions*.[71] And, over time, Leeuwenhoek abandoned his initial reservations about investigating semen, spending the next 20 years attempting to discover the secrets of seminal fluid. He in fact became involved in a very public dispute with no other than Martin Lister about sperm, pre-formation, and spontaneous generation. "The discovery of the future animal in one form or another within the spermatozoon would have been, after all, the crowning achievement of Leeuwenhoek's microscopy."[72]

However, we note that Leeuwenhoek stressed that he did not collect his sperm by masturbation, but within conjugal coitus. This is not surprising, as "medical and theological discourses were mutually dependent and influenced each other during this period… discussions about fertility in medical texts were implicitly and explicitly confined within the context of marriage."[73] Even when pushing boundaries of self-disclosure about his own body in the interest of natural philosophy, Leeuwenhoek stayed within parameters when he and his wife's social reputations were at stake.

We see the same sorts of parameters in the medical case of renowned natural historian John Ray. On 5 June 1673, Ray married the governess Margaret Oakley, a girl about half of his age. Ray was evidently a nervous groom, cognisant of the large age difference between he and his wife, "his austere outlook and frugal habits"

---

[67] Ibid, 217. See also "Letter from Leeuwenhoek to Lord Brouncker, November 1677," in van Leeuwenhoek 1941, vol. 2, 276, letter no. 35. Leeuwenhoek noted "I remember that some 3 or 4 years ago I examined seminal fluid at the request of the late Mr. Oldenburg."

[68] Ruestow 2004, 217. Oldenburg made his request on 24 April 1674.

[69] "Letter from Leeuwenhoek to Lord Brouncker, November 1677," in van Leeuwenhoek 1941, vol. 2, 276, letter no. 35. "Little earth-nuts" is evidently used here for 'common earth-nuts', *Bunium flexuosum Bith.*

[70] Ibid.

[71] van Leeuwenhoek 1677–1678, 1040–1046.

[72] Ruestow 1983, 210.

[73] Evans 2015, 17.

making it difficult for him to "adjust himself easily to the intimacies of marriage."[74] Since 1670, he had been discussing aphrodisiacs with Martin Lister by post. Ray had mentored Lister when Lister was a young naturalist, the two virtuosi sharing interests in ornithology, botany, and fossils; Lister contributed several locality records to Ray's plant and bird catalogues.[75] Theirs was a close and trusting friendship. In his letter of 13 February 1670, Ray wrote to Lister:

> Your opinion grounded upon experience, that *Opium* and all *opiates* are highly venereal, if taken moderately, I willingly embrace. The reason which induced me to subscribe to the common opinion, was not because I imagined them to be cold, as the former physicians fondly conceited, but because they do in a large dose fix the spirits and inhabit their motion, as appears in that they are anodynous[76] and soporiferous; and the sperm being a spiritous body, I was thence induced to think that they might hinder its turgescency. But, upon further consideration and inquiry, I find the effects of *Opium* to be something analogous to those of wine and other generous liquors, which, moderately drunk, incite to venery, but to excess, become soporiferous and narcotic, extinguishing that appetite. When it is supposed to proceed, that the Germans are of all nations most continent and least addicted to women.[77]

Early modern medical texts confirm that aphrodisiacs were a widely understood category of medicines, and opiates had been considered an aphrodisiac since Ancient Greece times as they decrease inhibition as well as sensation, which can lead to prolonged erections without ejaculation. This second effect would increase the possibility for the woman's orgasm, and could be perceived by both partners as a sexual enhancement. Endorphins produced by attachment to one's partner are also chemically very similar to opiates.[78]

In another series of letters that were "lost" or perhaps destroyed by William Derham, the editor of the first edition of Ray's correspondence, the short abstracts that remain reveal that in two letters written in July 1673, Lister advised Ray as soon as he was married to take diasatyria and cantharides, well-known aphrodisiacs.[79] One letter of 18 July 1673, however, escaped Derham's censorship, as it was amongst Lister's letters in the Bodleian Library.[80]

---

[74] Raven 2009, 177.

[75] Ray 2011, 19, 20, 34, 67–69.

[76] 'Anodynous' is a substance that alleviates pain or soothes.

[77] John Ray to Martin Lister, 13 February 1669/70, MS Ray 1, fol. 9, letter 25, Natural History Museum, London, printed Roos 2015, vol. 1248–53, letter 132.

[78] Miller 1992, 128; Faraone 1999, 126; Taberner 1985, 196–198.

[79] MS Ray 1, Derham Abstracts, f. 15, Natural History Museum, London. These substances are discussed as cures for sexual dysfunction in Pechey 1707, 176; Bonet 1684, 546. Bonet claimed that 'if the right great Toe be anointed with Oyl in which Cantharides have been dissolved it will cause an admirable erection." The Reverend William Derham (1657–1735), antiquarian, natural philosophy, and clergyman, succeeded Ray's close associate Samuel Dale (1659–1759) as the custodian of Ray's manuscripts. Derham adhered to the standards of his day, crossing out sections he considered irrelevant to include in red pencil, and discarding, keeping, or passing on letters of a personal nature.

[80] MS Lister 3, f. 241, Bodleian Library, University of Oxford. Printed in Roos 2015, vol. 1, 576–579, letter 255.

Ray opened his letter to Lister in English:

> I thank you most heartily for your kindness, exprest in so speedily answering my last, and for the encouragement you give me, and the good advice your Letter imports, which I doe highly approve of and intend to observe to follow. But because possibly you may not sufficiently understand my present condition, I shall now more fully open it to you.

And, then he switched to Latin and Greek as the nature of his plight became more sensitive in nature. He continued [my translation]:

> I think that I am suffering from a two-fold illness, for although in order to complete intercourse it is required first on the part of the male member both that it should become distended and erect, and also should remain hard and extended for a sufficient space of time, yet, if any erection does occur as a result of intense anticipation or an embrace, it is incomplete and momentary. Secondly, as far as sperm is concerned, it is required that it should not to be ejaculated without an erection at the first onset of sexual anticipation. But when I am in an embrace and a state of intense anticipation I can scarcely prevent my seed from leaping out without an erection. I cannot see how either of these problems is to be cured. As far as the first is concerned, since an erection occurs when the blood flows into and distends the spongy matter of the sinewy organs, and is kept there to prevent it quickly flowing back either by some contraction of the muscles or by some sinew which constricts the vein which conducts the blood back, then, if the reflux might be stopped or hindered using some technique for strengthening those muscles or that sinew, or for compressing the vein, it could be swollen and distended by means of straps as and when one wished. And yet (as far as I am aware) no method by which this might happen has ever previously been made known. For this reason I have no expectation of any assistance from any other source than some medicine of proven efficacy, of such great power and strength that it might bring about an erection, even if this were not readily forthcoming, and stiffness, irrespective of the manner in which it might eventually achieve it. There is however something which keeps me from total despair, because when I awoke after the first period of sleep following a lavish dinner before my wedding, or had been kept awake from eating some flatulent foods, then sometimes I felt a powerful stiffening. Since then this occurs at the prompting of nature, why might it not also be achieved by some artificial means? As far as the unstoppable outpouring of semen is concerned, it might perhaps be curbed by astringent medicines. You can see what a mess my affairs are in. I beg you to consider what needs to be done.
>
> I should be very happy to travel to York with my wife to enjoy your most delightful company and put into practice your advice for the present situation. For I have long been amazed by your instinctive good sense and cleverness in investigating the causes of things, and your success in what you discover. And if it is possible for me to be helped by medical assistance, then, with God's blessing, it is from you that I most particularly hope for relief and help. Yet my affairs are in such a bad way that I fear that it will be impossible this summer for me to undertake such a lengthy journey, and therefore I must consult you by letter.[81] I pray for good health and every blessing for you, your dearest wife and all your family. Farewell. Your most devoted and obliged friend.[82]

---

[81] See Raven 2009, 177.

[82] *Ego me duplici morbo laborare exsitimo nam cum ad coitum complendum requiratur primo ex parte Membri Virilis non tantum ut distendatur et erigatur sed ut sufficienti temporis spatio rigidus et extensus permaneat, mihi siqua erectio a forti imaginatione aut complexu succedat imperfecta ea est et momentanea. Secundo ex parte spermatis, non sine erectione ad primam rei venereae imaginationem ejiciatur. Ego a complexu et imaginatione forti vix contineo τὴν γονὴν ne absque erectione exiliat. Utrique huic malo quomodo medendum sit non video. Quod ad primum attinet, cum erectio fiat a sanguine fungosam corporum nervosorum substantiam influente et distendente,*

Ray was obviously suffering from premature ejaculation, his description of the problem in Latin and Greek resulting from no doubt personal embarrassment, and desire for privacy. In the Galenic tradition, as men were thought to be the hotter sex, they were thought to be the most capable sexually; the "suggestion that a man could lose his sexual drive had the potential to damage male patriarchal dominance and physical perfection."[83] As previous to his marriage Ray had been a celibate Cambridge don, he may have also worried as men who were subject to premature ejaculation were "considered effeminate, particularly sensitive individuals inclined to melancholy and the intense introspection brought on by an excess of studies."[84] He also worried about being able to provide for his young wife; until he inherited his mother's home in Black Notley, Ray had no fixed home for the first 6 years of his marriage, which made it difficult for him to visit Lister or to reciprocate hospitality.

Ray first indicated he would like to visit Lister with "my wife to enjoy your most delightful company and put into practice your advice for the present situation." As Evans has pointed out, the use of aphrodisiacs to treat infertility does not appear to have been fundamentally "sexed," seed provokers thought to be equally relevant for both men and women, so Lister's medicaments would be applicable to both.[85] Ray then noted that he was encouraged as he had an erection when flatulent. "Potent seed was thought to be filled with spirits and wind", a belief that can be traced to the eleventh-century treatise of Constantine the African who "noted that ingredients that contained nourishment and a cause of flatulence reproduced the essence of semen… Laevinus Lemnius also argued that the male organs needed to be filled with a "force of a flatulent spirit, whereby the seed may be driven forth" into the womb.[86] Wind thus helped with successful ejaculation of semen by maintaining an

---

ibique cohibito ne mox effluat vel per musculorum contractionem vel per nervum quendam venam reductricem stringentem: si refluxus arte sisti vel impediri possit vel musculos illos vel nervum roborando, vel venam comprimendo, peris tunc ad libitum inflari et distendi possit. Verum methodus qua haec fiant a nemine hactenus (quod sciam) prodita est. Quocirca ego juvamen non aliunde expecto quam a medicamento aliquo empirico adeo potenti et valido ut erectionem etiam invitam aut tentiginem inducat quocumque tandem modo id efficiat. Est autem quod facit ne prorsus desperem, quod ante nuptias a cena lautiore aut flatulentorum esu post primum somnum cum vigilaveram fortem tensionem aliquotiens senserim. Cum ergo hoc sponte naturae succedat cur non et arte procurari possit? Quod ad importunum seminis effluxum attinet possit fortasse is medicamentis astringentibus cohiberi. Vides quam in angusto sint res meae. Tu quaeso quod factopus est consulas. Ego quidem libentissime Eboracum una cum uxore proficisci vellem ut consuetudine tua iucundissima frui et consilio pro re nata uti pos-sem. Tuam enim jamdiu in rerum causis indagandis sagacitatem et δειvότητα, in inveniendis felicitatem miratus sum. et si medicis sub- sidiis mihi potest subveniri a te maxime favente numine leva-men et auxilium expecto. Verum eo in laevo sunt res meae, ut vereeor ne mihi hac aestate impossibile est tam longum iter suscipere per literas ergo mihi potius consulandus es. Tibi, conjugi dulcissimae tuisque omnibus prosperam valetudinem faustaque omnia precor. Vale. Tibi devotissimus et devinctissimus.

[83] Evans 2015, 81.

[84] Wilson 1993, 37

[85] Evans 2015, 88.

[86] Ibid., 109.

erection, creating an explosive force of movement, both things thought to promote conception as the semen "lost none of its potency."[87] Ray also noted that he experienced an erection after eating rich food, and indeed luxury food in early modern medical texts was "described as prompting and sustaining sexual activity."[88] Although we do not have Lister's full reply, as William Derham abstracted it, and then probably destroyed the original letter, we do know by September 1673, Lister was "comforting" Ray in his predicament, continuing to prescribe diasatyria. The medicament may not have worked very well; it did after all take 11 years of marriage before Ray's first child was born.[89]

On the other hand when he wrote Lister in 1673, Ray had some reason to hope. As the body and soul, were still imagined as one reality and could therefore not be separated, sexual dysfunction was not considered a solely emotional problem but was a tangible bodily condition that could be solved through employment of the appropriate aphrodisiac.[90] Thus Ray employed his keen empirical abilities, with a reasoned and detailed description. Although he was in despair, his affairs "a mess," he treated his condition as a scientific and medical problem to be solved. Although Ray did not publish his own predicament due to its sensitive nature, like Wilkins and Lister, he was his own case study.

10.4 Conclusion

In the late seventeenth-century, the generation of public datasets overcame previous mindsets about bodily privacy during illness. These conceptual shifts were due to developing norms for the formulation and publication of early modern medical and scientific case studies. There was increasing objectification of the body, self-experimentation, an increasing precision of empirical description inspired by Baconian models, and the subjection of bodily fluids to chemical experimentation. Virginia Woolf noted that when confronted with indescribable pain, the physician was "forced to coin words himself, and taking his pain in the hand, and a lump of pure sound in the other. .. to crush them together that a brand new word in the end drops out." But she may have been unduly pessimistic when she thought the result would probably be "laughable"; rather, early modern physicians and natural philosophers seeking the language to describe the body and its pains began to find their voice, beginning with themselves and their own experiences with disease and dysfunction.

[87] Ibid., 109

[88] Gowing 1996, 91.

[89] MS Ray 1, Derham Abstracts, 15, Natural History Museum, London, printed in Roos 2015, 562, letter 262.

[90] Evans 2015, 81.

Bibliography

Manuscripts

Bodl = Bodleian Library, University of Oxford
RS = Royal Society Library, London

Primary and Secondary Sources

Anonymous [Martin Lister]. 1672. An account of a stone cut out from under the tongue of a man. *Philosophical Transactions* 7: 4062–4064.

Baynard, Edward. 1695. An account of the probably causes of the pain in rheumatisms. *Philosophical Transactions* 19: 19–20.

Beukers, Harm. 1999. Acid spirits and alkaline salts: The iatrochemistry of Franciscus dele Boë, Sylvius. *Sartoniana* 12: 39–59.

Bigotti, Fabrizio, and David Taylor. 2017. The Pulsilogium of Santorio: New light on technology and measurement in early modern medicine. *Society and Politics* 11 (2/22): 55–114.

Birch, Thomas. 1756–1757. *History of the Royal Society*, 4 vols. London: Printed for A. Miller in the Strand.

Bonet, Theophile. 1684. *Mercurius compitalitius: A guide to the practical physician*. London: Thomas Flesher.

Boss, Jeffrey. 1978. The medical philosophy of Francis Bacon (1561–1626). *Medical Hypotheses* 4 (3): 208–220.

Churchill, Wendy. 2013. *Female patients in early modern Britain: Gender, diagnosis, and treatment*. Aldershot: Ashgate.

Cook, Harold. 1996a. Natural history and seventeenth-century Dutch and English medicine. In *The task of healing: Medicine, religion and gender in England and the Netherlands, 1450–1800*, ed. Hilary Marland and Margaret Pelling, 253–270. Rotterdam: Erasmus Publishing.

———. 1996b. Physicians and natural history. In *Cultures of natural history*, ed. Nicholas Jardine, James Secord, and Emma Spary, 91–105. Cambridge: Cambridge University Press.

Descartes, René. 1998. *The world and other writings*. Trans and ed. Stephen Gaukroger. Cambridge: Cambridge University Press.

Duden, Barbara. 1991. *The woman beneath the skin: A doctor's patients in eighteenth-century Germany*. Cambridge: Harvard University Press.

Ecklund, Jon. 1975. *The incompleat chymist: Being an essay on the eighteenth century chemist in his laboratory, with a dictionary of obsolete chemical terms of the period*, Smithsonian studies in history and technology, number 33. Washington, DC: Smithsonian Institute Press.

Evans, Jennifer. 2015. *Aphrodisiacs, fertility and medicine in early modern England*. Martlesham, Suffolk: Boydell and Brewer.

Faraone, Christopher A. 1999. *Ancient Greek love magic*. Cambridge: Harvard University Press.

Grew, Nehemiah. 1675. *Disputatio medico-physica, inauguralis, de liquore nervosa… pro gradu doctorates… subjicit Nehemiah Grew, e. com Warwickensi, die 14 Julii.*, Ph.D. Diss. University of Leiden.

Gowing, Laura. 1996. *Domestic dangers: Women, words and sex in early modern London*. Oxford: Oxford University Press.

Hall, Marie Boas. 1999. *Henry Oldenburg: Shaping the Royal Society*. Oxford: Oxford University Press.

Herr, Henry W. 2008. 'Cutting for the stone': The ancient art of lithotomy. *British Journal of Urology* 101 (10): 1214–1216.

Hooke, Robert. 1935. In *Diary of Robert Hooke, 1672–80*, ed. Henry W. Robinson and Walter Adams. London: Taylor and Francis.

Jardine, Lisa. 2004. *The curious life of Robert Hooke: The man who measured London*. London: Harper Collins.

Keller, Eve. 2003. The subject of touch: Medical authority in early modern Midwifery. In *Sensible flesh: On touch in early modern culture*, ed. Elizabeth D. Harvey, 62–81. Philadelphia: University of Pennsylvania Press.

Kronick, David A. 1988. Anonymity and identity: Editorial policy in the early scientific journal. *The Library Quarterly* 58 (3): 221–236.

Lamont-Brown, Raymond. 2001. *Royal Poxes and potions: The lives of court physicians, surgeons, and apothecaries*. Stroud, Gloucestershire, Sutton: Sutton Publishers Ltd.

Lieberman, Matthew D., et al. 2007. Putting feelings into words: Affect labeling disrupts amygdala activity in response to affective stimuli. *Psychological Science* 18 (5): 421–428.

Lindemann, Mary. 2010. *Medicine and society in early modern Europe*. Cambridge: Cambridge University Press.

Lister, Martin. 1683. *Letters and divers other mixt discourses*. York: John White.

———. 1684. *De Fontibus Medicatis Angliae, Exeritatio Altera*. London: W. Kettilby.

———. 1685. An account of a stone grown to an Iron Bodkin in the bladder of a boy: Communicated by Dr. Lister Fellow of the Royal Society. *Philosophical Transactions (1683–1775)* 15: 882.

———. 1711. *Dissertatio de Humoribus*. London: Gul. Bowyer.

Miller, Richard Alan. 1992. *The magical and ritual use of aphrodisiacs*. Rochester: Inner Traditions Bear and Company.

Oldenburg, Henry. 1968–1971. *The correspondence of Henry Oldenburg*. Trans. and ed. A.R. Hall and M. Boas Hall, vols. V–VIII. Madison: University of Wisconsin Press.

Pechey, John. 1707. *The compleat herbal of physical plants*. London: R. and J. Bonwicke.

Pomata, Giannetta. 2005. *Praxis historialis*: The uses of *historia* in early modern medicine. In *Historia: Empiricism and erudition in early modern Europe*, ed. Giannetta Pomata and Nancy Siraisi, 105–146. Cambridge: MIT Press.

Ragland, Evan. 2012. Chymistry and taste in the seventeenth century: Franciscus Dele Boë. *Ambix* 59: 1–21.

Raven, Charles. 2009. *John Ray, naturalist: His life and works*. Cambridge: Cambridge University Press.

Ray, John. 2011. *John Ray's Cambridge catalogue*. Trans. and ed. by P.H. Oswald and C.D. Preston. London: The Ray Society.

Riha, O. 1995. Surgical case records as an historical source: Limits and perspectives. *Social History of Medicine* 8: 271–281.

Roos, Anna Marie. 2008. All that glitters: Fool's gold in the early-modern era. *Endeavour* 32 (4): 147–151.

———. 2011. *Web of nature: Martin Lister, the first scientific arachnologist*. Leiden: Brill.

———. 2015. *The correspondence of Dr Martin Lister (1639–1712)*, Vol. 1. Leiden: Brill.

Ruestow, Edward G. 1983. Images and ideas: Leeuwenhoek's perception of the spermatozoa. *Journal of the History of Biology* 16: 185–224.

———. 2004. *The microscope in the Dutch Republic: The shaping of discovery*. Cambridge: Cambridge University Press.

Sandassie, S. 2008. Evidence-based medicine? Patient case studies in English surgical treatises, 1660–1700. *Medical Humanities* 34: 11–18.

Scarry, Elaine. 1985. *The body in pain: The making and unmaking of the world*. Oxford: Oxford University Press.

Shapiro, Barbara. 1969. *John Wilkins, 1614–1672, an intellectual biography*. Berkeley: University of California Press.

Shelley, Harry S. 1958. Cutting for the stone. *Journal of the History of Medicine and Allied Sciences* xiii (1/2): 50–67.

Siena, Kevin P. 1998. Pollution, promiscuity, and the pox: English venereology and the early modern medical discourse on social and sexual danger. *Journal of the History of Sexuality* 8 (4): 553–574.

Simpson, W. 1669. *Hydrologia Chymica*. London: Printed by W.G. for Richard Chiswel.

Skloot, Rebecca. 2010. *The immortal life of Henrietta lacks*. New York: Crown Archetype.

Slare, Frederick. 1683. A discourse about the Calculus Humanus. *Philosophical Transactions* 14 (157): 523.

Stallybrass, Peter. 1986. Patriarchal territories: The body enclosed. In *Rewriting the renaissance: The discourse of sexual difference in early modern Europe*, ed. M. Ferguson et al., 123–142. Chicago: University of Chicago Press.

Taberner, P.V. 1985. *Aphrodisiacs: The science and the myth*. Philadelphia: University of Pennsylvania Press.

Tomalin, Claire. 2002. *Samuel Pepys: The unequalled self*. New York: Knopf.

van Helmont, Johannes. 1648. *Opuscula medica inaudita: I. De lithiasi. II. De febribus. III. De humoribus Galeni. IV. De peste*, reprinted Brussels, 1966. Amsterdam: Elzevier.

van Leeuwenhoek, Anthony. 1677–1678. Observationes D. Anthonii Lewenhoeck, De Natis E Semine Genitali Animalculis. *Philosophical Transactions* 12: 1040–1046.

van Leeuwenhoek, Anthoni. 1941. *Alle de brieven. Volume 2: 1676–1679*. Amsterdam: Swets & Zeitlinger.

Webb, W.W.rev. Scott Mandelbrote. 2004. Sherard, James (1666–1738). In *Oxford dictionary of national biography*. Oxford, Oxford University Press.

Wilson, Lindsay. 1993. *Women and medicine in the French enlightenment*. Baltimore: Johns Hopkins University Press.

Wittie, Robert. 1660. *Scarbrough Spa*. London: Charles Tyus.

Woolf, Virginia. 2012. *On Being Ill*, with introductions by Hermione Lee and Mark Hussey, afterword by Rita Charon. Ashfield: Paris Press.

Index

G. Manning (ed.), *Testimonies: States of Mind and States of the Body in the Early Modern Period*, Archimedes 57,
https://doi.org/10.1007/978-3-030-39375-5

CPSIA information can be obtained
at www.ICGtesting.com
Printed in the USA
LVHW082252020720
659625LV00013B/277